William Falconer

An Account of the Efficacy of the Aqua Mephitica Alkalina

solution of fixed alkaline salt, saturated with fixible air, in calculous disorders, and

other complaints of the urinary passages. Fourth Edition

William Falconer

An Account of the Efficacy of the Aqua Mephitica Alkalina
solution of fixed alkaline salt, saturated with fixible air, in calculous disorders, and other complaints of the urinary passages. Fourth Edition

ISBN/EAN: 9783337100629

Printed in Europe, USA, Canada, Australia, Japan

Cover: Foto ©berggeist007 / pixelio.de

More available books at **www.hansebooks.com**

AN

ACCOUNT

OF THE EFFICACY OF THE

AQUA MEPHITICA ALKALINA;

OR,

SOLUTION OF FIXED ALKALINE SALT,

SATURATED WITH FIXIBLE AIR,

IN

CALCULOUS DISORDERS,

AND OTHER
COMPLAINTS OF THE URINARY PASSAGES.

BY

WILLIAM FALCONER, M.D. F.R.S.

AND

PHYSICIAN TO THE GENERAL HOSPITAL AT BATH.

THE FOURTH EDITION:

With Additions, Alterations, and several new and remarkable Cases, not inserted in any former Edition.

LONDON

PRINTED FOR T. CADELL, IN THE STRAND;

AND SOLD ALSO BY J. KILLICK, NO. 7, BROAD WAY, BLACK FRIARS, NEAR LUDGATE-HILL.

M DCC XCII.

[Price THREE SHILLINGS.]

TO

BENJAMIN COLBORNE, Esq.

OF THE

CITY of BATH.

DEAR SIR,

PERMIT me to congratulate you on the increasing reputation of a remedy, whose efficacy you have experienced so fully, and to which your present happy state of health, as well as that of several of your friends and acquaintance, is most undeniably to be ascribed.

The benefits to be imparted to mankind by its publication, have, I know, been your only motives for wishing the information, which these sheets may contain, to be as widely dispersed as possible, in order that by variety of communications its true character

The honour, however, of the difcovery of what I apprehend to be its moft important quality hitherto known, that of relieving calculous complaints, is due to a gentleman of this city, Benjamin Colborne, Efq. who had formerly been of the medical profeffion, which he practifed many years with great reputation to himfelf, and fervice to mankind.

Having been a fevere fufferer from this diforder, he was induced to make trial of feveral of the moft celebrated remedies, but was, after long and fad experience, convinced but too well of the inefficacy or danger of moft, if not all of the fo boafted lithontriptics. He was then led, fortunately for himfelf, to make trial of the remedy now under confideration; and the event anfwered much beyond his hopes, and has added greatly to his happinefs; not only by the relief he has himfelf experienced from it, but alfo by the opportunities it has afforded him of indulging, in the moft difinterefted manner, his benevolent difpofition, by recommending its ufe to feveral of his friends who laboured under the fame malady.

Mr. Colborne was led to this difcovery, partly from obferving the diffolvent powers
of

of alkaline falts upon the urinary calculus out of the body, and ftill more by remarking the changes produced by their internal ufe on the urine of thofe afflicted with thefe diforders, rendering that clear and of a natural colour, which was before turbid and difpofed to precipitation. The difagreeable tafte, however, of the uncombined alkali, which is moreover fo naufeating to the ftomach, together with its cauftic, feptic, and irritating effects on the animal fyftem, the urinary paffages particularly, were great difcouragements to its ufe. Could thefe be obviated by any combination that would ftill leave the alkaline falt at liberty to unite with the acid that is fuppofed to contribute to the formation of thefe calculi, the purpofe of preventing their being generated, or poffibly of diffolving them when formed, would probably be in a good meafure anfwered.

Fixible Air feemed to him adapted to this purpofe in every refpect, as it forms with the alkali a neutral falt, perfectly mild in its nature, agreeable to the tafte and ftomach, and powerfully antifeptic. At the fame time their combination is fo loofe, that the alkali is eafily feparated from the air by any other acid it may meet with.

He moreover found by experience, that this combination poffeffed no inconfiderable diffolvent powers upon the human calculus out of the body. Hence he was induced to make trial of it himfelf, and to recommend it to others. The accounts of its fuccefs here fubjoined will, I truft, prove that his expectations were not ill founded.

The moft convenient method of preparing the alkaline folution is as follows. Put two ounces and a half troy weight, or, if troy weights are not at hand, two ounces and three quarters * avoirdupois, of dry falt of tartar into an open earthen veffel, and pour thereon five full quarts, wine meafure, of the fofteft water, that is clean and limpid, that can be procured, and ftir them well together with a clean piece of wood. After ftanding 24 hours, carefully decant, from any
indiffoluble

---

* Two ounces and a half troy weight contain 1200 grains; two ounces and three quarters avoirdupois contain 1201 grains and a quarter of a grain.

indiffoluble refiduum that may remain, as much as will fill the middle part of one of the glafs machines for impregnating water with Fixible Air *. The alkaline liquor is then to be expofed to a ftream of Air according to the directions commonly given for impregnating water † with that fluid. When the alkaline folution has remained in this fituation till the Fixible Air ceafes to rife, a frefh quantity of the fermenting materials fhould be put into the lower part of the machine, and the folution expofed to a fecond ftream of Air, and this procefs repeated four times. When the alkaline liquor fhall have continued about 48 hours in this fituation, it will be fit for ufe, and fhould then be carefully drawn off into perfectly clean bottles (pints are I think preferable), and clofely corked up. The bottles fhould then be placed

---

* If the falt of tartar be good, and perfectly foluble in the water, every ounce meafure of the alkaline folution fhould contain feven grains and a half of alkaline falt.

† The directions given with the machines fold by Mr. Parker in Fleet-ftreet, or by Meffrs. Neale and Bailey, No. 8, St. Paul's Church-Yard, will be fufficient for thofe who choofe to prepare this remedy themfelves.

placed with their bottom upwards * in a cool place; and with thefe precautions it will keep feveral weeks, and perhaps much longer, very good. The quantity of alkaline folution above directed to be mixed at the beginning of the foregoing directions, is judged to be fufficient to fill the glafs machines of the common fize twice over, without pouring off the liquor fo deep as to hazard making the folution turbid, by ftirring up the indiffoluble refiduum which is precipitated at the bottom † of the veffel. The water in which the alkali is diffolved, fhould be as free of foreign impregnations as pofiible, as the alkali, by decompofing them, will not only cloud the water, but form other combinations inconfiftent, perhaps, with the effects to be wifhed for from the remedy. The intention therefore of mixing the falt of tartar with the water the day before, and of the caution recommended in pouring it off, is to allow time

* A fhelf with holes in it to put the necks of the bottles into, fuch as are commonly ufed for wine decanters, is convenient for this purpofe.

† If the alkaline folution as above directed fhould be found too irritating, it may be made with a fmaller proportion of the falt. See Mr. Melmoth's Cafe.

time for any precipitation occafioned by the mixture to fettle, as well as to feparate the indiffoluble parts of the falt of tartar itfelf. Nor is lefs attention neceffary in procuring the falt of tartar pure and in perfection; and on that account it fhould be got from fuch places only as can be depended upon. When properly prepared, the alkaline mephitic water fhould be perfectly clear and rather fparkling, of an acidulous tafte, and totally free of that difagreeable impreffion which alkaline falts make on the tongue and throat [*].

About eight ounces by meafure appear, from fome of the cafes, to have been taken thrice in 24 hours for a confiderable time together, and to have agreed well with the ftomach, appetite, and general health; but I apprehend moft people will think this too large a quantity; and I believe that, for moft cafes,

---

[*] Thofe who do not choofe, or to whom it is inconvenient to prepare this remedy themfelves, may have it, made according to the above directions, of ohn Killick, No. 7, Broad-way, Blackfriars, near Ludgate-Hill. For the conveniency of carriage, as well as the better prefervation of the virtues of the preparation, it is put into fmall bottles, each of which contains one moderate dofe.

cafes, two thirds, or a pint of the alkaline liquor in 24 hours, may fuffice: fhould the bulk of the feparate dofes * be thought too large, the alkaline folution may be made of double the ftrength; in which cafe, half the quantity will be enough. The times of taking three dofes in the day have been, I believe, pretty early in the morning, about noon, and about fix in the evening. If twice a day, about noon and in the evening; and if once, which in many cafes feems fufficient for a preventive, about an hour and a half before dinner. Common prudence dictates that fuch a remedy fhould be taken at fuch times as the ftomach is leaft likely to be loaded with victuals.

I do not find, from obfervation or inquiry, that a rigid adherence to any particular regimen of diet is neceffary, farther than the ufual prudential cautions of moderation and temperance.

The Rev. Dr. Cooper has made ufe of fruit, wine, and other things fubject to acetcency,

See Dr. Cooper's Cafe.

acefcency, during the time of his taking the folution; yet no perfon, as will appear by his very judicious account, has received greater benefit. I, however, think it would be advifable to abftain from acids, and from fuch things as are fubject to become acefcent, for fome time before, and alfo after the time of taking the dofes of the alkaline folution. I do not find, either from my own obfervation, or from the accounts of others, that any very perceivable effects, fave that moft to be wifhed, the abatement of the troublefome fymptoms, followed the taking this remedy. I have inquired of a very fenfible perfon of this city, who has taken the folution in the largeft quantity of any that I have known; and he affures me, that he found no effect from it, fave that of gently opening the body.

Mr. Bewly fpeaks of a dofe of it that he took affecting the head (with vertigo I fuppofe), and proving a pretty ftrong diuretic. But fuch confequences have not been obferved by all thofe perfons of whom I have had an opportunity of inquiring. The perfon before referred to, informed me, that though it kept

the

the body gently open, it had no effect in increasing the quantity of urine. Mr. Bewly's dose was indeed large, he having taken, at one dose, such a quantity of the alkaline solution as contained 24 ounces by measure of Fixible Air, whereas the quantity of air taken at a time in a dose of the solution above directed, is not calculated to exceed 15 ounces; but this was repeated three times a day, and no such effect observed. With respect to the diuretic quality, it is well known that the expectation of such an effect from any thing we take, will often prove a very powerful means of producing it.

Should it prove cold or flatulent to the stomach, as I have myself known it to do, though I believe that rarely happens, a small portion of spirits, as rum or brandy*, or any of the other spirituous waters or tinctures, may be used without any diminution of its good effects. A tea-spoonful of rum is mentioned to be taken with each dose of the solution, in one of the cases subjoined; and I have myself directed a small quantity of tincture of cardamoms and of compound spirit of lavender, with

See Dr. Cooper's Case.

with evident advantage. Mr. Colborne has found hot milk, in the proportion of about one fourth to that of the alkaline folution, to be a very grateful addition, efpecially in cold weather, and what tended much to reconcile it to the ftomach, and this without impairing in the leaft its good qualities.

If the urinary paffages are very fore or tender, and the fyftem very irritable, it will be neceffary to ufe opiates. Five, ten, twenty, or thirty drops of tincture of opium, or a proportionable quantity of the paregoric elixir, muft be taken as neceffity may require, once or twice a day during the reft of this remedy. The opiate may be taken either juft before, or juft after the taking the alkaline water; but the quantity of the opiate fhould be diminifhed gradually, and at laft totally laid afide, when the pain and other urgent fymptoms have either ceafed, or fo far abated, as not to caufe any great uneafinefs.

CASE

CASE I.*

Benjamin Colborne, Efq. of this city, was, in the year 1760, attacked with a violent nephritic paroxyfm, which, after continuing feven or eight days, and being treated with anodyne, oily, and mucilaginous medicines and bleeding, terminated in the difcharge, by urine, of a red ftone larger than a vetch or tare, after which he continued tolerably well for eight or ten months; often, however, obferving fmall calculous concretions to come away, attended with irritation of the urinary paffages. In about ten months after the firft attack, he had another, but neither fo violent or of fo long duration, which terminated like the firft, in the difcharge of a ftone of a fimilar colour to the foregoing, but of a fmaller fize. The nephritic paroxyfm again returned in about five or fix months, but not fo violent as at firft. During this time he was

* The Cafes bef re related, are here reprinted as in the third Edition. What has been added in the two laft editions is put down in *Italics*, with the date prefix-d. Where no addition is made to the Cafes before printed, no information has been received of the patient's health.

was in a courſe of taking mucilaginous and lubricating remedies.

After this he made trial of Mrs. Stephen's remedy, as prepared by Dr. D'Eſchernay, of which he took about an ounce in a day, once or twice a week.

After this he continued free of nephritic complaints about a year and a half. That medicine, however, agreed ſo ill with his ſtomach, producing nauſea, indigeſtion, and crudities, that he was obliged to leave it off. About three or four months afterwards he had another attack, which returned again upon him every ten or twelve weeks. At this time he was in a courſe of taking an infuſion of the wild carrot ſeed, and drank diſtilled water as his uſual drink.

In the year 1766, he made a trial of Blackrie's lixivium (or Chittick's remedy); and though it agreed with him rather better than the ſoap, yet it was ſo cauſtic and irritating to the mouth and throat, and produced ſuch painful ſenſations in his ſtomach, that he was obliged to leave it off; after which, his

his nephritic paroxyfm returned every eight or ten weeks as before. In the year 1774, he went to Spa for a complaint in his bowels, which he afcribed to the ufe of his cauftic lixivium, and, during the time of his drinking thefe waters, had no return of calculous complaints; but on his coming back to England he was attacked as formerly.

In the beginning of the year 1778, he made trial of water fimply impregnated with Fixible Air, which proved too irritating and diuretic. On March 27th of the fame year, he had an attack of the gout, which continued on him until the 14th of April, when he was taken with a violent vomiting, attended with pain in the left kidney. By the help of the warm bath and bleeding, he paffed another calculus. After this he had a fecond attack of the gout, which continued a few days.

As foon as it was over, he began the ufe of the alkaline medicine with Fixible Air, as above defcribed, of which he took about fix or feven ounces twice a day. During the ufe of this he parted with no gravel, his urine depofited no fediment whatfoever, or difco-

loured

loured the veffel, though, if it was omitted even for a few days, thefe appearances took place, and fmall bits of gravel were perceivable in his water.

From this time he continued in perfect health, and free of all nephritic complaints, until the 26th of Auguft, 1783, when, about three in the morning, he was taken with an irritation in the urinary paffages, which prevented his fleep; his urine however was not high coloured: about feven in the morning he had two purging ftools; he had but little pain in the kidney, but a heavy obtufe fenfation over the os pubis, which continued with fome ficknefs till about two o'clock, when the ftone feemed to enter the bladder. From that time he became perfectly eafy.

In order to difcharge the ftone from the bladder, he drank large quantities of mucilaginous liquors, and retained his urine as long as poffible. About fix in the evening he difcharged a red calculus, fmaller than what he had before done.

It

It is proper to obferve, that he had been at Harrowgate about four or five weeks before this happened, and drank the Harrowgate water, which as it acted not only as a purgative, but as a diuretic alfo, he was induced to think he might fafely omit the alkaline folution. It appeared however, to his great difappointment, that the calculus was generated during that interval. From that time to the prefent, he has never, for two days fucceffively, omitted taking the mephitic alkaline folution, and has never fince felt the fmalleft uneafinefs; no grains of fand or other precipitation in the urine, nor any difcolouration of the veffel, except when the medicine is omitted for a day. But, upon taking the folution again, the urine made afterwards diffolves the former difcolouration, and ftill continues perfectly clear. During the time he was fubject to nephritic paroxyfms, his urine was fubject to putrify very foon; but fince he has taken the folution, it will keep three or four days in the warmeft weather without fhewing any figns of that difpofition. His general dofe as a preventive is about feven ounces daily. His health, ftrength, and fpirits, are all perfectly good; and, as

he

he thinks, better than they were twenty years ago.

Since the above account was written, which is now about two years ago, Mr. Colborne has had two fits of the gout; the one flight, the other more severe, which laft confined him for a fortnight; both fits, however, went off perfectly well, without any tranflation of the gout to the head, lungs, or any of the vifcera. He drank the mephitic alkaline water, with the addition of a little brandy, during both the fits, and it agreed with him perfectly well. His health, ftrength, and fpirits, are as good now as they were two years ago.

December 16, 1788.

Mr. Colborne has taken the mephitic alkaline water but once or twice in a week for four or five months paft; yet his chamber-pot has kept clean from any incruftation or adhefions. He thinks that his appetite has been better upon the days he took the mephitic alkaline water. He has had more of the gout this November (1788) *than for three or four years paft.*

December 1, 1791.

Mr. Colborne informs me, that he has had no return of his complaints for many years paſt, notwithſtanding his having often omitted taking the alkaline water for three or four months together; yet, even during that time, his urine seldom forms any depoſit that adheres to the chamber-pot.

CASE II.

Mrs. Southcote, a lady of this city, was firſt afflicted with complaints of this kind about the year 1754, when ſhe had an attack that laſted ſeveral days; after which, to her great ſurpriſe, ſhe voided a calculus, not having before apprehended the nature of her diſorder. She continued free from any complaint of the kind for about ten years, when, in the year 1764, ſhe had a return, and from that time the attacks recurred every ten or twelve weeks, accompanied with the diſcharge of numerous calculi: one, however, ſhe had reaſon to believe remained, and probably ſtill remains too large to paſs, which aggravated her pain, and produced blood on the ſlighteſt motion.

motion. The pains in the kidneys, nevertheless, still continued; and the last paroxysm she had of this kind, which was in 1779, was so violent, that her life was despaired of for ten or twelve days. At last, however, after taking large quantities of oily and mucilaginous remedies, the free use of opium, and the warm bath, an oblong stone was discharged, about the shape and size of a large orange seed. As soon as she had recovered a little strength after this severe attack, she began, in the same year, to make trial of the aqua mephitica alkalina, taking six or seven ounces twice a day, which she still persevered in. Since that time she has had no more nephritic attacks, has parted with no calculi, and her urine continues clear and free of sediment. She often feels a sensation of weight, and some uneasiness in her bladder, but has any bloody water, bears exercise well, is able to walk for an hour or two at a time, and uses a carriage almost daily without pain or aggravation of uneasiness. Her general state of health, though valetudinary, has been much better since the trial of this remedy than before.

Since the above case was drawn up, and sent to the printer, I have received the following account:

"In the beginning of September, 1784, Mrs. S. went into Berkshire, where she continued three weeks. Soon after her arrival she was seized with a feverish complaint, which occasioned her to omit the alkaline solution, which she not only discontinued during her stay in the country, but for a month after her return to Bath. During this interval, she began to feel some uneasiness in her left kidney, on which she again commenced the use of the remedy once a day. On Dec. 27th last, she was seized with pain and other symptoms attending the passage of a calculus; which, after a painful night, came away about nine the next morning. It was about the size of a pea. She soon recovered, and is now well and easy."

It is her opinion, and appears highly probable, that this calculus was formed during the time of the remedy's being omitted. Mrs. Southcote has had no return of her nephritic complaint, and is able to walk and bear the motion of a carriage without the least pain or inconveniency; but feels sometimes an uneasy sensation in the bladder, and believes she has two calculi formed there. April 30, 1787.

Since

Since the above account was published, Mrs. Southcote had two attacks of the apoplexy, the last of which carried her off, after a short illness, on January 1st, 1788, aged 68 years. Her body was opened by Mr. Symons, an eminent surgeon of this place. Her liver was found in a putrid state; the gall bladder of the size of a hen's egg, and its coats a full quarter of an inch thick. In the middle of the cystic duct was lodged a biliary calculus, of the size of a child's marble, which might be pressed back into the gall bladder, but not forwards. The bladder and kidneys were perfectly sound, and free from any calculous concretion of any kind.

CASE III.

The Reverend Dr. Cooper, of Sunning, in Berkshire, a most worthy and amiable character, is likewise a remarkable instance of the efficacy of the neutralized alkaline solution.

But this gentleman's case is related by himself, in a letter to my late friend, with such accuracy and propriety, as well as animated description, suggested by the memory of feel-

ings too severe to be erased, that I cannot forbear giving it to the reader in his own words; subjoining also a confirmation of the benefit he had received, and of his present good state of health, extracted from a letter I myself had the pleasure lately to receive from him. One trivial circumstance I will take the liberty to remark, that Dr. Cooper, in one part of his letter to Dr. Dobson, seems to have thought that the aqua mephitica alkalina, or alkaline solution saturated with Fixible Air, was recommended in the Medical Commentary, as a remedy for calculous disorders; whereas it is only recommended there as containing a large quantity of Fixible Air, which was to be set loose by a subsequent addition of an acid, which was directed to be taken immediately after the exhibition of the alkaline solution. It does not appear that Dr. Dobson, at the time he composed the Medical Commentary on Fixed Air, was at all acquainted with the good effects of the alkaline solution thus impregnated in these complaints. Though he recommends its use, it is only with a view to its immediate decomposition by an acid. The remedy, however, from which Dr. Cooper received benefit, was, as appears from his own account, the alkaline
solution

solution saturated with Fixed Air, without any other addition; though he occasionally made use of the effervescent saline draught, when a machine for impregnating the water with Fixed Air was not at hand.

The Rev. Dr. Cooper's *Letter to* Dr. Dobson.

" Dear Sir,

" It gives me great pleasure to hear you design taking up the pen again in favour of *Fixed Air*. The efficacy of that volatile principle (when combined with some alkaline salt) in putrid and other disorders, is sufficiently manifested in your very ingenious Commentary on that subject; and nothing now is wanting completely to establish its character, than the making better known to the world its superior virtues in nephritic complaints also. Of this superiority, I am sensible, you have several proofs before you, even in this place, and some of them much stronger than perhaps my case may be; nevertheless, if *that* can in the least degree promote the cause of truth, and assist your benevolent design, it is most heartily at your service. Indeed, I feel myself under so great obligations

obligations to the virtues of *Salt of Tartar and Fixed Air*, for rescuing me from a state of misery and pain, and restoring me to the full enjoyment of health and ease, that it would appear the highest ingratitude in me to be silent, whenever it is in my power to do justice to their worth.—It was in the beginning of August, 1772, if I recollect right, that I was first attacked with what is called a fit of the gravel, which lasted about twelve hours; *hinc mihi prima labes*. As I had till then been quite a stranger to the nature as well as symptoms of the disorder, I was at a loss how to account for the sickness and pains I felt, till a small stone, which came away, too well convinced me from what cause they arose. The continual apprehensions I now was under, of having a return of those pains, and the dread I entertained of being afflicted with a complaint which I had always heard styled the *opprobrium medicorum*, destroyed every comfort, and embittered every hour of my life. I did not fail, however, you may be sure, Sir, having recourse to the best advice I could find, and took care scrupulously to adhere to every rule and every method of cure prescribed me. I soon perceived, nevertheless,

theless, with great concern, that my disorder, instead of abating, gradually increased, conformable to the just observation of Mr. Pope, that

> "The young disease, which must subdue at length,
> "Grows with our growth, and strengthens with
> " our strength."

I now continually voided great quantities of sand, or rather, of very small stones of a bright red colour; and, at the distance of every two or three months, and sometimes oftener, when a larger stone was formed than could easily pass the ureters, I underwent the most excruciating torments before it reached the bladder. The paroxysms, at these times, lasted full thirty hours, and once or twice much longer, attended with an acute burning pain in the region of the kidneys and round the abdomen, a numbness down my thighs and legs, a constipation of my bowels, with violent sickness at my stomach. Castor-oil, fomentations, emollients, and warm bathing, which used before to afford me ease in common fits, here often failed of success, and nothing but opiate draughts could administer the least relief. Nor did my sufferings always terminate with the stone's being at length safely
lodged

lodged in the bladder; for twice, in its endeavours to pass the urethra, the stone unhappily remained fixed there for several hours, and consequently brought on again an intolerable pain, with a total suppression of urine. To attempt giving an idea of what I felt on those occasions, is beyond the power of words; even at this distance of time, while I am now writing, *animus meminisse horret*—it is to be conceived only by those who have had the misfortune to be afflicted with the stone.

" As I was convinced that the milder remedies, which I had hitherto followed, were unable to prevent a frequent return of these paroxysms, I determined to have recourse to more violent ones, such as *lixiviums* and *solvents*. Of the former, I preferred that recommended by Mr. *Blackrie*, known before by the name of *Chittick's Receipt for the Stone*. This I took regularly for four months, strictly observing the rules laid down with it. I do not remember I had any very violent attack of my complaint, during the course of this medicine; but it sometimes occasioned me to make bloody water, and I continually voided a good deal of gravel. Perceiving, however,

however, that my health, spirits, and appetite began to be afflicted by the septic regimen, enjoined to assist the operations of the lixivium, I thought it high time to leave it off; and soon after had the additional mortification to know, that, whilst every thing else, that could render life an object of desire, was about to leave me, my calculous complaints remained firm and rooted as ever.

" From this caustic medicine, I turned my eyes to *Perry's Solvent*, which, as I found its character and virtues came strongly recommended to the public under the sanction of many respectable names, I lamented I had not thought of sooner, and considered all the time as thrown away, which I had hitherto bestowed on other remedies. My application, however, to this boasted medicine, was followed by no better success than what had attended me before; for at the end of three months, during which time I took it, I found all my fond hopes and expectations at once destroyed by one of the severest fits of the stone I had ever felt. Willing to give this celebrated solvent the fairest trial, I persevered in the use of it long after I found it by

no

no means suited to my conftitution; for it induced such a coftive habit of body, as rendered my life very uncomfortable, and sometimes was indeed quite alarming.

"It would be difficult, as well as tirefome, to endeavour to enumerate the variety of other noftrums, which, during the courfe of full feven years, I was perfuaded to fwallow:

"Non, mihi fi linguæ centum fint, oraque centum,
"Ferrea vox, omnes poffim comprendere formas."

Let it fuffice to fay, that finding from none of them any other kind of benefit than temporary fufpenfions of pain, I quite defpaired of ever meeting with any thing that would afford me effential and permanent relief. At length, however, in the beginning of April 1780, a friend of mine put into my hands your publication, before mentioned, on Fixed Air: pleafed with the account given in it, of the many cures performed by *that* and *falt of tartar*, on putrid and other difeafes, and with the great probability of the fuccefs of thefe combined articles in nephritic complaints, as likewife encouraged by the eftablifhed character and reputation of its amiable author, I determined immediately to make trial of this extraordinary

extraordinary medicine; and accordingly provided myself with a Fixed Air machine, and apparatus neceffary for the purpofe.

" About the middle of the fame month I entered on a courfe of the Medicated Water and Fixed Air, taking it in the form and quantity prefcribed as in your pamphlet, and foon had great reafon to congratulate myfelf on my undertaking; for in about a fortnight's time I perceived a very fenfible alteration in myfelf, as well with refpect to my complaint in particular, as to my health in general. The latter I found greatly mended both in my fpirits and appetite; and the uneafy fenfations of the former, about the kidneys, were entirely removed. I no longer voided either fand or gravel; nor did I feel that continual irritation to make water, which I did before; nor was my fleep difturbed by fuch frequent, yet fruitlefs, calls to it: in fhort, from the happy enjoyment of eafe and comfort, to which I had fo long been a ftranger, I now feemed to myfelf quite a new creature.

" I purfued this method about four months, when my farther progrefs in it was
ftopped

stopped by a feverish attack, which confined me for three weeks. As soon as that was removed, I had recourse again to the *Salt of Tartar* and *Fixed Air*, and have continued it, with but little interruption, ever since. I can assure you, Sir, with the greatest truth, that from the time I began taking this medicine, to the date of the present writing, I have never had any the least return of my complaint, excepting once, about two years and a half ago, I voided a small stone, without pain, about the size of a little pea, or vetch, quite smooth, and almost perfectly round. I have moreover, in every other respect, enjoyed an uninterrupted state of good health. When I am on a journey, or absent from home, when I cannot be supplied with a Fixed Air machine, I neutralize each dose of the medicated water (sweetened with a little sugar) with juice of lemons, before I take it, which has the same effect as the mephitic acid.

" With regard to regimen, I confess, I observe none, except the avoiding every thing salted or dressed too high. No other restriction of diet can be necessary with a medicine, whose virtues seem best assisted by those things which are, at the same time,

most

most salutary and agreeable to the nature and constitution of the human frame, such as wine, milk, fruits, vegetables, and the like. On this account, the medicine in question has certainly great advantages over those of the caustic kind; for the same reason, perhaps, it may be supposed to yield to them in *solvent powers*. Nevertheless, if, as experience shews, it *prevents* the *formation* of those substances in the kidneys and bladder, which form the *human calculi*, or the *increase* of them after they are formed, its claim to merit as a *preventative* is equally great; at the same time, when its perfect innocence, nay, even beneficial effects on the constitution, are taken into consideration, few people, I believe, will hesitate to pronounce the *Medicated Water* and *Fixed Air* superior to all other medicines hitherto recommended for nephritic complaints. A fair trial of them for three years, will, I hope, fully justify me in asserting this superiority; and if health, ease, and comfort, are blessings we all covet and desire, the having reinstated me in the happy enjoyment of them, when well-nigh lost, must ever entitle the *Salt of Tartar* and *Fixed Air* remedy to my
sincerest

sincerest thanks and most grateful acknowledgments.

  I am, dear Sir, with the truest esteem,
    Your very obedient,
      and very humble servant,
Bath, April 16th,   EDWARD COOPER.
1783.

"N. B. I forgot to mention, that, in the spring of the year 1782, I was seized with a fit of the gout in both my feet, which confined me full three weeks; neverthelefs, I still continued the use of my medicine, adding only to each dose about half a tea-spoonful of rum; nor did I find the least prejudice or inconvenience from it."

Extract of a Letter from the Rev. Dr. Cooper *to* William Falconer, M. D. *dated* Dec. 18, 1784.

"All that I have farther to add now, respecting myself, is, that I still continue as well, and as free from any return of my complaint, as I was when I drew up my case in April 1783. I constantly persevere in the use of the alkaline solution with Fixible Air, drinking once or twice a day, as it happens,
            about

about two ounces of the Medicated Water, which never fails acting as a preventative, and keeping me intirely free from every the leaft fymptom of gravel or ftone; though I have great reafon to think, from the pain I have formerly felt in the region of my kidneys, that a ftone is formed in one of them."—

It may be neceffary to remind the reader, that the quantity of alkali contained in the folution ufed by Dr. Cooper, is double to that ufed by Mr. Colborne; fo that the two ounces mentioned in Dr. Cooper's letter as his daily dofe, are equal to twice that quantity of the folution directed in the former part of this Work.

Extract of a Letter, dated April 1, 1787.

" My health is, I thank God, full as good as when you laft heard from me, nor have I had any, even the leaft, return of my old complaint, which I can attribute to nothing but my perfevering in a courfe of the above-mentioned medicine. I purfue the fame method of taking it that I have done hitherto, excepting that now I have feldom recourfe to it above once a day, inſtead of twice, which I formerly had. The effects of it as a preventative I find equally powerful as they have ever been; and long experience has fully convinced me that this medicine is no leſs innocent, than it is ſalutary to my conſtitution. It is true indeed I have had two fits of the gout, the one in September 1785, and the other in January 1787, but in neither did I deſiſt from taking the Aqua Mephitica Alkalina. On the contrary, I continued it during the whole time of both thoſe ſevere viſitations, and, with the precaution of adding a teaſpoonful of rum or brandy to each doſe, found it ſit quite eaſy and comfortable on my ſtomach."

The following is an Extract from a Letter I received from Dr. Cooper, *dated* November 26, 1788.

"*To anſwer more particularly your inquiries concerning my preſent ſtate of health, I have the pleaſure to aſſure you that it is full as good as when I ſaw you laſt winter at Bath, 'bating the ſomewhat farther advance in age, and its attendant infirmities. I believe I then told you,*
that

that for the whole preceding Summer, and great part of the Autumn, I had been afflicted with a very painful and dangerous illness; and that during my confinement under it, which lasted near five months, I was obliged to abstain from the Aqua Mephitica Alkalina. I found however no inconvenience from the disuse of it, either by any return of pain in my kidneys, or any other (even the least) hint of a gravelly complaint. Since that time, I have again entered upon a course of that medicine, but neither in so large a quantity, nor so frequently repeated as before, as I now take it once only in the day, and that not regularly. Besides, whenever business or engagements call me from home, I oftentimes omit it for a fortnight together, and find myself justified in this omission by a total freedom from every symptom of, or tendency to, either the stone or gravel. Should I unfortunately find any hints of either of these sufficient to alarm me, I should immediately increase my dose, and be more attentive to the frequent and regular returns of taking it; nor have I the least doubt, but that the virtues of the Aqua Mephitica Alkalina would soon disperse every anxious fear and uneasy foreboding."

December 1, 1791.

Dr. Cooper has had no return of his nephritic complaint, though he does not take the Mephitic Alkaline Water constantly.

CASE IV.

A respectable person of this city, who desired his name might not be made public, aged 65, of a habit of body esteemed to be scorbutic, had been for several years accustomed to the use of medicines that acted upon the urinary organs, as expressed juice of millepedes and tincture of cantharides.

About three years ago he was seized with a considerable degree of pain in the urinary passages, and in the rectum. He likewise voided several fabulous concretions, some of the size of a pepper-corn, or vetch, and had frequent returns of bloody urine, in which the proportion of blood was often so large as to coagulate nearly in the same manner as if it was recently drawn from the arm. Great pain, as may well be supposed, attended these evacuations. For these symptoms he took, by advice, Blackrie's

Blackrie's lixivium, from forty to eighty drops, thrice a day, in veal broth or onion pottage, and made a large ufe of onions alfo in his diet.

His pains and bloody urine increafing under this regimen, he was induced to make trial of honey, which he took to the quantity of near half a pound daily, ftill continuing the ufe of the lixivium. The honey feemed to act as a ftrong diuretic, and to aggravate his pain fo much, as to render it neceffary to be laid afide, as well as the lixivium. He then made trial of water, fimply impregnated with Fixible Air, for about a month, but without any fenfible relief.

He next, by Mr. Colborne's advice, entered upon a courfe of the Alkaline folution impregnated with Fixible Air, fimilar to that above defcribed, which he commenced fomewhat more than two years from the prefent time, taking eight ounces of it thrice every day. In lefs than three weeks after his firft taking it he experienced the moft effential benefit; his pains abated, his urine became clear and of a natural colour, without any fubfidence or precipitation; and his health (fome flight pains,

occa-

occafionally returning, excepted) nearly reſtored.

It is proper to remark, that the cauſtic lixivium appeared to have very bad effects on the fyſtem, by difpoſing the humours of the body to a putrefactive ſtate, which was inſtanced in feveral refpects, and particularly by frequent hæmorrhages from the nofe, that occurred during its ufe; a thing he was never before fubject to, and which has not occurred ſince the lixivium has been laid afide.

For the laſt year and half he has made no bloody urine, has had no pain in paſſing it, and has voided no calculous concretions. For the laſt fix months he has taken only four ounces three times a day, which is but half the original quantity. It has operated as a gentle aperient, giving one motion daily, but no more, and thus fupplying the want of an aloetic pill, which he was formerly obliged occafionally to have recourfe to. It had not, however, any fenſible effect as a diuretic.

His appetite and health in general have been very good ſince the ufe of the medicine. He is now of a healthy and ruddy complexion, hale and strong in his body, appetite and spirits good, bears exercise well in a carriage, and is able to walk five or six miles at a time without fatigue, or any other inconvenience, and generally walks as far daily, whenever the weather will admit of it. I had the satisfaction this day (April 14, 1787) of seeing the person whose case is here described. He is in perfect health, and strong and active for his years, and has had no return of any calculous complaints these two years. He has continued, and still continues to take regularly every day, the Mephitic Alkaline Water.

The person whose case is here related, continues to take the Mephitic Alkaline Water occasionally, but not constantly; and is quite free from his former complaints of the calculous kind.

November 25, 1791.

The person here referred to, is, I am informed by his family, free from any calculous complaint at present; but has had one or two attacks, which were but very slight. He has however been far from regular in the use of the remedy, having often omitted it for a considerable time together.

CASE V.

The Honourable and Reverend G. Hamilton, of Taplow, in Buckinghamshire, a gentleman between 60 and 70 years old, is another instance of the efficacy of this remedy; as appears from the following extract from a letter of his to George Burges, Esq. of this city, and by him communicated, with the consent of Mr. Hamilton, to Dr. Dobson.

" I had been troubled with a stone in my bladder about five years, during which time I took various solvents without any effect. In the spring of the year 1780, Mr. Pott extracted a stone weighing two ounces; since that time I have been free from pain, but at times perceived gravel in my water, and now-and-then pieces large enough to make me apprehend the forming of another stone. In the winter of the year 1781, I was at Bath, and very fortunately became acquainted with Dr. Cooper. He had been troubled with my complaints, and was taking a medicine he strongly recommended to me. He said, he had taken it near two years, to the best of my remembrance,

brance, during which time he had avoided the usual symptoms of this complaint. It was water impregnated with Fixed Air, to two quarts of which he put two ounces of salt of tartar. He took a small quantity of this twice a day, in which he put some sugar, and about two tea-spoonfuls of juice of lemons. He very kindly treated me with a glass whenever I called upon him; and as soon as I returned to Taplow, I sent to town for a Fixed Air apparatus. I got it in January 1782, and immediately entered on the course prescribed by Dr. Cooper: only that I drink his two doses at once, and put the juice of half a lemon into mine, as my stomach agrees well with acids. Since I have taken this, I have voided no gravel; nor have I seen any fur on the chamber-pot, its usual forerunner.

"If this account may be of any service to Dr. Dobson, or his patients, he is welcome to make what use he pleases of it; for I may say with Dido,

"Haud ignara mali," &c.

Dated Taplow,
Apr. 8, 1783.

The following is a Copy of a Letter I lately had the pleasure of receiving from Mr. Hamilton.

S I R, Taplow, April 10, 1787.

"YOUR letter of the 7th reached me this morning; in anfwer to which I can inform you, that I continue taking the Fixed Air and Salt of Tartar, and think I find benefit from it. I this winter had occafion to confult Dr. Warren for fome complaints, the chief of which was lofs of appetite. He advifed me to leave off my medicine whilft I was taking his prefcriptions; I did fo for fome weeks, till I found fome ugly pains in my back, occafioned by fome very fmall ftones. This alarmed me, and made me return again to my Fixed Air, which foon relieved me. I recommended it fome time ago to a Mr. Wood, my hofier in Piccadilly, and to Mr. Charteris at Eton, and they both affure me they have found great benefit from it. I forgot to mention that I voided three of the fmall ftones above mentioned. If what I have faid will be of any fervice, you are welcome to infert

my

my letter in the new publication you have in hand.

I am, Sir,
Your obedient humble fervant,
G. HAMILTON."

Mr. Hamilton is fince dead; but, as I hear, had no return of any calculous complaint.

CASE VI.

William Ainflie, Efq. of this city, a gentleman between 80 and 90 years of age, accuftomed to take much exercife, in hunting particularly, was feized in the year 1780 with a pain and irritation in the urinary paffes, accompanied with a difcharge of blood. This continued eight or nine days, but without his paffing any gravel or fand.

He continued tolerably well (though not without frequent irritations of no great confequence, in which, however, nothing of a calculous nature was voided) until Auguft 1781, when he was again feized with violent pain and irritation, accompanied with bloody urine, which came on after exercife on horfeback. After fome time his urine became clear,

clear, but a violent irritation remained for two or three days longer; nothing, however, of ſtone or gravel came away.

In January 1782, he came from Dorſetſhire to Bath in a chaiſe, the motion of which renewed his complaints, the irritation particularly, to ſuch a degree, as to make it difficult for him to reach the end of his journey. Soon after his arrival at Bath, he was adviſed to a trial of Adams's ſolvent, of which he took ſomewhat more than a guinea bottle; during the taking of which he thought himſelf ſomewhat better, the irritation being rather diminiſhed: but towards the latter end of February he was ſeized with a great bleeding at the noſe, which continued 48 hours, with the loſs of ſome quarts of blood.

The phyſician he conſulted on this occaſion adviſed him to leave off the medicine; but from that time the leaſt motion brought on pain, irritation, and bloody urine, ſo as to oblige him to ſtir out as little as poſſible, ſince even the motion of a ſedan-chair brought on the above ſymptoms.

In this ſtate he continued, although he was ſtill in the courſe of taking various mild
lubricating

lubricating things, and laudanum occafionally, to abate the pain, until about the beginning of April 1783, when he was advifed to begin a courfe of the Alkaline Solution faturated with Fixed Air, which he took to the quantity of eight ounces twice a day. He had not taken it more than five or fix days, before he found benefit: his pain abated, he became able to walk a little; but much motion ftill brought on a return of bloody urine, and the other fymptoms, but lefs in degree, and of a fhorter continuance than formerly.

By the beginning of May he was fo well recovered, as to venture to take a journey in a poft-chaife into Dorfetfhire. The firft day of his journey he travelled about 25 miles; and the roads being very rough, and the carriage uneafy, brought on a return of his pain and bloody urine. He however went forward about 15 miles the next day, and the roads being better, and the carriage eafy, felt no inconvenience. The next day brought him about 25 miles farther, to the end of his journey, where he arrived in perfect eafe and health.

About the twentieth of May he began to diminifh the quantity of his medicine, taking
it

it once a day only. From this time he remained perfectly well until the beginning of August, when he had a very flight return of pain, which foon ceafed. In October following he was able to ride a horfe gently for an hour and a half together without much pain or uneafinefs; and his water was then, and had been a long time, of a natural colour, plentiful in quantity, and voided without pain or uneafinefs.

Mr. Ainflie died on the fifth of May 1786, aged 87 years, of a peripneumonic complaint contracted by expofure to cold, but without any fymptoms that could be afcribed to calculus, or any diforder of the urinary paffages.

CASE VII.

Mr. John Rolfe, of Amefbury, in Wiltfhire, was attacked in May 1779 with a fit of the gravel, which lafted near two months, and was at times very painful, and attended with ficknefs, vomiting, and lofs of appetite, which continued until a calculus as large as a vetch came away. He then grew better, and recovered his appetite. Between the time

time above mentioned, and the year 1784, he had frequent returns of the same complaint, which caused sickness and vomiting in him for a day, and sometimes for two or three days before the gravel came away, which sometimes would be nearly as large as a barley-corn. He drank on these occasions an infusion of marsh-mallow roots, with gum arabic, and other mild softening ingredients, and took castor-oil occasionally. On Saturday Nov. 13, 1784, after being warmed with hunting, he became on a sudden cold and chilly, with aching pains similar to those of the rheumatism in his sides, breast, back, and limbs, which continued several days, and seemed rather to increase. On the 22d of the same month he felt some pain in his stomach and bowels, and applied on that occasion to his apothecary, who gave him some opening physic, which operated very properly downwards, but caused sickness and vomiting. This induced him to think his disease to be of the bilious kind; and in consequence thereof he took some remedies which gave him some relief, but did not remove the pains in his back, breast, &c. Having at that time some business in Dorsetshire, he thought the journey might be of service to

his

his health; and on Thurſday the firſt of December he went to Saliſbury, and from thence to his friend's houſe in Dorſetſhire, which was about thirty miles diſtant from the laſt-mentioned city. This journey, which he performed in one day on horſeback, increaſed his pain ſo much, that it was with difficulty he got to the end of his journey. The next day he was better, but not free from the pains above mentioned; his appetite alſo was very indifferent, his body coſtive, and his ſtools hard and black. In this ſtate, nearly, he continued until the tenth of December, on the evening of which day he was ſeized with ſickneſs at his ſtomach, and a vomiting of blood in conſiderable quantity: the night following he was again ſeized with the vomiting of blood, on which Dr. Pultney, of Blandford, was ſent for, who recommended to him ſome medicines, that ſtopped the bleeding. He continued at his friend's houſe until the laſt day of December, when he returned home, with his pains much as before, and his health very indifferent. After he had been at home about eight or ten days, he paſſed a ſtone much larger than any he had before done, being of the ſhape of a barley-corn, but larger. After this he found his health and
appetite

appetite better. After a few days, however, his stomach became again painful, sore, and tender, and often subject to vomiting; and it was with great difficulty he could bear the motion of a post-chaise. His complaint being now judged to be a combination of bile, rheumatism, and gravel, he was ordered to Bath, where he arrived May 5, 1785. Dr. Adair was sent for a few days after his arrival, who directed him some medicines, and a cautious trial of the Bath waters. His health however not improving, his medicines were altered, and a blister applied to the stomach, which soon became easier in respect of pain, but his other pains in the back, breast, &c. remained as before, and his sickness and vomiting continued. Dr. Falconer was consulted about this time, who, in conjunction with Dr. Adair, advised him to try the Alkaline Water with Fixible Air, to be taken in the quantity of a pint a day, divided into three doses, and to use the warm bath twice a week. He had not taken the Alkaline Water above two or three times, before he found his appetite mended, and his pains rather abated. After continuing this course for three weeks, his stomach became settled, his appetite returned, his sickness left him

him by degrees, and his vomiting ceased entirely; and from that time he continued in perfect health, and free of all gravelly complaints. During the last illness that he had, his urine was of a deep colour; and notwithstanding he made but a small quantity in the course of a night, it left a great sediment in the vessel. But from the time he began to drink the Alkaline Water, his urine became pale, came away freely, and in large quantity, and left not the least fur or discolouration on the vessel. *The above account is taken from a letter of* Mr. Rolfe *to me, and expressed as nearly as possible in his own words. The letter is dated from* Amesbury, *October* 19, 1785.

I have, since the time above specified, had the pleasure of the following account from Mr. Rolfe, *dated* Amesbury, *January* 30, 1787:

" I should sooner have given you a line respecting my health, had it been otherwise than well. It is at present so good, that I do not remember it to be better; though I have voided a stone in the course of the last summer, in a decayed and crumbling state,

as

as it fell to pieces by the preffure of my fingers. I account for my having this ftone by my having in April laft broken my glafs apparatus, and my not being able to get another for the fpace of a month; and in this interval I apprehend the ftone was generated. I ftill continue the Alkaline Water impregnated with Fixible Air, taking about one third of a pint three times a day."

CASE VIII.

Lieutenant-colonel Gould, aged 55, formerly of the third regiment of guards, was fubject to gouty attacks at little more than thirty years of age, which have, at intervals, attacked him ever fince. About nineteen years ago, he was taken with frequent ftoppages of urine, that would fometimes continue three or four hours, and were not relieved but by the ufe of a bougie, which he was obliged frequently to have recourfe to, and once even to wear one conftantly for two months together. Thefe fits at firft had long intervals, but of late years have become more frequent. His urgings to make water were fo frequent, that the retaining of it was very difficult, and often impracticable. His urine was moftly turbid,

turbid, and had a ſtrong tendency to putrefy, and had an evident fœtor of that kind when voided. He uſed alſo to paſs a large quantity of mucus in his urine, inſomuch that it would ſtand nearly one inch deep in the pot, when there was no more than a pint and a half of water. He has at times paſſed red gravel, but never any of ſuch a ſize as could be called a calculus. He had conſiderable pain and ſenſation of weight in the hypochondria, which ſometimes extended to the back. At the time of making water he had an inclination to go to ſtool. When the gout was in the extremities, he was generally free from theſe complaints. About March laſt, 1785, he made trial of a vegetable diet of rice milk, maſhed potatoes, turneps, &c. and drank no fermented liquor, but barley-water only. After continuing this regimen about three weeks, he was ſeized with a ſmarting pain in the urethra, and paſſed about half a pint of blood daily for two days ſucceſſively. Opiates, ſpermaceti draughts, with a change of diet to one of animal food, and the moderate uſe of wine, relieved theſe ſymptoms, which were followed by a ſlight fit of the gout. During the ſummer, air, and moderate exerciſe on horſeback daily for ſix weeks, amended his

general

general health, but his urine ſtill continued turbid and fœtid. About the latter end of Auguſt laſt, he had a ſlight and favourable fit of the gout, which laſted about three weeks; his urine, however, continued as before. On Friday, October 7th, he began, by Mr. Colborne's advice, to take the Aqua Mephitica Alkalina, of which he drinks about one third of a pint twice a day: he takes it with a little hot milk, and it agrees perfectly with his ſtomach. His hours of taking, are about two in the afternoon, and about ſix in the evening. It tends to keep the body regular as to ſtools, but has no farther purgative effects. It has ſhewn no ſtimulating effects upon the urinary paſſages, as his wants in that reſpect are much leſs frequent than formerly. In five days time, after he firſt took it, his urine began to grow clear, and void of mucus or fœtor, and to be eaſily retained. He can now keep it three or four hours, and paſſes it, though not without pain, with much leſs than he formerly did. It is of a moderately pale amber-colour, void of ſmell, and will keep twenty-four hours, and probably would much longer, without ſuffering any change, and leaves no fur on the chamber-pot. His health and appetite

petite are much better since he took this remedy, and his strength and ability to walk much improved. His regimen of life has been fish or plain meat, with half a pint of red port wine, or perhaps more, and it has agreed well with him. He has taken fruit at times, but thought it disagreed with him.

The above account was taken down from Colonel Gould's own mouth in the month of January 1786. But although the Mephitic Alkaline Water gave great relief from pain by abating the acrimony of the urine; yet it appeared that the bladder and urinary passages, and indeed the constitution itself, were so injured and weakened by so many repeated attacks in a long course of years, as to be irremediable. He died of an internal mortification at the latter end of 1786.

CASE IX.

Mr. Francis Loftus, of Market-Weighton, in Yorkshire, a person of sixty-seven years of age, and good constitution and general state of health, except with regard to this particular complaint, had been twice cut for the stone, the last of which operations was performed

formed somewhat more than eight years previous to the writing of his first letter to Benjamin Colborne, Esq. which bears date Jan. 27, 1786. He there mentions that he is satisfied that another is formed in his bladder; and though he does not there mention the symptoms that induced him to be of that opinion, it may reasonably be presumed that he must, from sad experience, be sufficiently qualified to decide upon such a question. Having seen in one of the monthly publications an account of the success of the Aqua Mephitica Alkalina, he made trial of it according to the receipt there put down, which by mistake directed two ounces of salt of tartar instead of one, to be dissolved in two quarts of water. This he tried for near six weeks to a pint and half daily, but without any abatement of his symptoms, save that his water, which was before turbid, and deposited a sediment that adhered to the vessel, became almost clear.

Mr. Colborne, however, having informed him of the mistake in the printing the receipt, and advised him to make trial of a solution of half the strength only, he in his next letter, dated March 7, 1786, gives a more

more favourable account. His pain in paffing his water was abated, and he was able to retain it longer. He adds in this letter, that the ftone was an hereditary complaint in his family, that his father had it, but did not live to be cut. His next letter, dated June 7, 1786, contains little more than an account of his farther amendment in general terms, and a confirmation from examination of his having a calculus formed in his bladder. His urine, he adds, is ftill rather fœtid. His next letter, dated July 25, 1786, gives a much more favourable account. He there defcribes his pains as having fubfided, his urine having loft its fœtor, and his health being perfectly reftored. He adds, that he could then walk three or four miles *with a great deal of eafe*. He had alfo left off his opiate, as being unneceffary. The Mephitic Alkaline Liquor, he fays, never purged him, but kept his body moderately open.

His next letter, dated October 6, 1786, confirms the opinion of the efficacy of the medicine. He there fays that his health is ftrong and good, and fuperior to what could be expected at his time of life, and that he had juft walked four miles without the leaft
inconve-

inconvenience. A subsequent letter, dated January 9, 1787, is to the same purpose. He expresses himself as being " in great good health and spirits, and surprisingly strong, quite free from pain, and able to walk three or four miles with pleasure." He adds, that he now takes the water only twice a day, and finds that to answer his purpose perfectly well.

In a Letter to Mr. Colborne, *dated* December 6, 1788, Mr. Loftus *expresses himself in the following manner:*

" *I have never neglected taking the Alkaline Water one day since you heard from me, but have diminished the dose to half a pint taken once a day. You will say then I am strong; and so I am, as I can walk four or five miles a day with pleasure, and can also ride on horseback. I never see any fragments in the pot; but something like small rags, and a red sharp sand sticks to the bottom and sides. I am wonderfully strong, and in good health, and am seldom troubled with any pain.*"

It appears, by a letter from Mr. *Loftus to* Mr. *Colborne, dated* May 17, 1789, *that his complaints*

complaints had returned, and that he was founded nine or ten times at York by a gentleman of eminence in the profession, but that no calculus could be discovered. It was, however, conjectured that he had an ulcer in the bladder, from the white sediment in his urine. In a second letter, dated June 8, 1789, he says, the white sediment in his urine is diminished, but that he is no easier: since that time I have heard no account. It is no wise remarkable that an ulcer of the bladder should happen after the operation of lithotomy being twice performed.

CASE X.

William Melmoth, Esq. a gentleman now at an advanced period of life, was, many years ago, subject to violent pains in the back on much walking, which he attributed to weakness induced by profuse discharges of blood by the bleeding piles; a complaint under which he had laboured several years. During the time of his being subject to the above-mentioned disorder, he once or twice, after exercise on horseback, made some coffee-coloured urine. About sixteen years ago, he had an attack of a gravelly complaint, accompanied with calculous discharges. This recurred

curred at longer or shorter intervals, and attended with more or less pain, until a period of about 15 months preceding the present time, when he first had recourse to the Mephitic Alkaline Water. Of this he took about half a pint daily, of the strength mentioned in the former part of this work, and persevered in this course for about a month or six weeks. During this space he always discharged with his urine pretty large and numerous particles of coagulated blood, but without any pain. This appearance caused him to suspect that the medicine operated upon the urinary passages with too great violence. On this presumption he gradually reduced the strength of the solution, and at last found that three drachms of salt of tartar, dissolved in two quarts of water, was the proportion that agreed best with him. Specks of blood are scarcely now ever to be seen in his urine. Of the preparation above mentioned he takes, and has for some months taken, about half a pint daily, and often a less quantity. Mr. Melmoth's health is much better now than it had been for several years previous to his trial of the above-mentioned remedy, and his strength much restored. He is also at present able to walk about the town without

exciting

exciting any gravelly fymptoms, which formerly were excited by very flight motion, infomuch that he could fcarcely ftir out of doors without ufing a fedan chair, which is now no longer neceffary on that account.

Mr. Melmoth has always taken, as well during the ufe of the Mephitic Alkaline Water as before, a moderate proportion of vegetables, and fuch other acid fubftances as are ufually eaten with animal food, and has ufed for drink at meals two or three glaffes of wine, and now and then a little ale or porter. He has never found the Mephitic Alkaline Water to difagree with his ftomach in any refpect. In very cold weather he fometimes puts into his cup a fingle tea-fpoonful of brandy.

December 14, 1788.

I this day received an account from Mr. Melmoth, in anfwer to an inquiry after his ftate of health, in which he fays, " that it continues in the fame good ftate it was when he gave the account inferted in the former edition; that he has continued to ufe the Mephitic Water prepared in the fame manner, and taken in the fame fmall quantities as before defcribed; and that,

that, since he first had recourse to it, he has never felt the slightest return of his complaint."

December 3, 1791.

I am informed by Mr. Colborne, that Mr. Melmoth continues perfectly free from any complaint of the urinary passages, and in a good state of health in other respects, notwithstanding his having omitted the use of the Alkaline Water for six or seven months together.

CASE XI.

Mr. Patrick Westoby, of Gainsborough, in Lincolnshire, a person now advanced in years, was, one day in the summer of 1776, seized with a sudden stoppage of urine, whilst he was passing his water freely. This continued a few minutes, and was not attended with much pain. To relieve this he took, as soon as it could be got, an infusion of Burdock, with the addition of some milk, and a little honey, which passed off freely in the night following by urine, and brought with it a small round calculus, flat on one side, and with a little rising on the other, and about 1-5th of an inch diameter. He then continued

tinued free from any complaint of this kind about a year, after which interval he was again attacked with frequent urgings to make water, which he paffed in fmall quantities at a time, and with fome, though but little pain. For thefe complaints he had recourfe again to the infufion of Burdock root, but did not find the fame eafe from it as before. That year and the following he took a confiderable quantity of Adams's Solvent, but did not experience any relief from it. For the two or three years next fucceeding, he ufually had, in each, two or three attacks; but they were moderate, and relieved by gentle diruretics with the addition of opiates. During the two years immediately preceding his laft attack, his paroxyfms were lefs frequent, but more urgent, and attended with more bloody water, and more frequent excitings to pafs his water, and greater pain in the urinary paffages, which continued to be very troublefome for fome nights, but went off in the fpace of two or three weeks. On the 3d of Febuary 1785, when the froft was very intenfe, he had a fmart attack, attended with many urgings to make water, which he did about thirty times in the courfe of the night, and paffed his urine tinged with blood. In a few days however thefe fymptoms abated,

and

AQUA MEPHITICA ALKALINA. 63

and his urine flowed freely and plentifully, and became of a good colour, and paffed without pain. During the time he was in bed he was fubject to a flow of pale clear water, which caufed fuch quick returns of inclination to pafs it, as interrupted his fleep very much. The bark in infufion, and opiates, removed this fymptom, and in about a month all his complaints ceafed.

In the month of Auguft following he had another attack, which continued about three weeks, and another towards the latter end of the month of October, which continued, with two intervals of 12 days each, to the 18th of January 1786. The weather was then very fevere, and his diforder never more troublefome; his urine bloody, with frequent urgings to pafs it, pain and irritation in the urinary paffages, and no fleep but with opiates. He began that day, about noon, to take the mephitic alkaline folution, and continued to take it regularly, according to the directions in the Appendix to Dr. Dobfon's work, for four five days, but without any perceptible alteration in his fymptoms, except that his urine became a little clearer. On Sunday, January 22, he paffed the day and night in great pain; but
on

on the Monday and Tuesday after, he observed a brown powder at the bottom of his chamber-pot, in quantity nearly sufficient to cover a shilling, which, he thought, proceeded from the dissolution of a calculus. He then became somewhat easier, but remarked, during two or three of the evenings preceding, that he felt some pain in the back, os pubis, and thighs, a little below the hips, all of which were new symptoms. From Wednesday morning his symptoms began to decline; his water passing freely, and being of a good colour, his irritations of the passages abating, and his urgings to pass it returning less frequently. During this time he observed in his urine what he took to be fragments of a calculus. Before he took the mephitic alkaline solution, he took a diuretic pill of oil of juniper, gum arabic, and uva ursi, during the use of which his urine was red, thick, and turbid, and deposited a sediment that adhered closely to the pot. His disorder continued to abate until February 7, 1786, when he had another attack, attended with pain and some bloody urine, which lasted a few days. His urine after this attack became again pale and clear; but nevertheless was voided in small quantities, and attended with some pain.

He

He then, by Mr. Colborne's advice, diminished the quantity of the solution that he took daily, and added a few drops of laudanum to each dose; which abated his sensations of irritation in the passages, and his urine soon became clear, and void of mucus or blood, was voided without pain, and easily retained. In June 1786, he was so far recovered as to be able to walk three miles a day without pain, feeling nothing more than a slight obtuse sensation at the time his last drop of urine came away. His urine deposited, on standing, a considerable quantity of a white adhesive sediment, and sometimes small bits like broken shells. From this time to April 1787, he went on taking the Mephitic Alkaline solution in about two-thirds or one-half of the dose directed in the Appendix, and has felt no return of his complaint, except once, which he ascribed to the breaking of the machine, which prevented his taking his remedy for about seven weeks, after which he had a moderate fit, attended with frequent excitements to pass urine, but not much pain. The Mephitic Alkaline Water being again procured, abated these symptoms, and carried them all off in about a fortnight. At present, April 4, 1787, he is free of all pain and irritation, or other ſymptoms of calculus.

CASE XII.

A ſervant of Charles Sheppard, Eſq. of Caſtle Godwin, near Painſwick, Gloucesterſhire, was, about the month of October 1785, taken with a ſuppreſſion of urine, and obliged to have recourſe to a catheter, in the paſſing of which a ſtone was diſcovered in his bladder. This neceſſity recurred two or three times during the winter; and the poor man, in conſequence of this complaint, was rendered incapable of labour, and even unable to walk a very few yards without making bloody urine. He had alſo a conſtant diſcharge of mucus by the urinary paſſages. In May 1786, he began to take the Mephitic Alkaline Water; and before he had taking it a fortnight, he found relief, and in about ſix weeks could walk or ride on horſeback without any diſcharge of blood, and the mucus diſcharge was much leſſened. During the courſe of laſt winter he had one or two attacks, which were apprehended to proceed from his drinking cyder; his ſtate of health is now eaſy and comfortable;

he has walked ten miles, and rode eighteen with much eafe; can work whole days in the garden, but is cautious of any violent exertion, as he fometimes feels the ftone as a weight, and has frequent calls to make water, and fometimes has a mucus fediment in his urine. *Taken from a letter of Charles Sheppard, Efq. dated Caftle Godwin, April* 11, 1787.

CASE XIII.

The Rev. Arthur Evans, about 50 years of age, and of a fcorbutic habit of body, has been for fome years afflicted with frequent nephritic complaints. In November 1785, he was feized, in a moft violent manner, with repeated vomitings and pains in the kidneys. In December following he came to Bath; and having the pleafure of Mr. B. Colborne's acquaintance, he confulted him, who kindly gave his advice, and recommended the immediate ufe of the Mephitic Alkaline Water, advifing three or fpoonfuls of boiling milk to be mixed in the water till the ftone had paffed; half a pint was drunk every day, according to this direction. In a week from the firft drinking, an oblong ftone of the fize of a small kidney bean was passed with little pain; it was three weeks and five days from the first seizure to the passing of the stone. Mr. Evans persevered in taking the same quantity of Mephitic Water daily (without milk) till June 1786, when he left Bath, and soon had the misfortune to break his machine, which deprived him of the Mephitic Water full three weeks; the consequence was, that towards the latter end of August he had a return of his nephritic complaint, but no vomiting: in a few days afterwards he voided a few calculi; when these had passed, he soon became easy, and remained perfectly free from that complaint till February last, when he was suddenly seized with frequent vomitings and pains in the kidneys: these nephritic symptoms came on in the evening, and early in the subsequent morning a small stone was passed, which Mr. Colborne, on examining, thought was rather a fragment of an old one than a new concretion; as Mr. Evans had not omitted drinking daily half a pint of Mepihtic Water from August to February last, from which last period Mr. E. has had no complaint of a nephritic nature.—The foregoing account is given in Mr. Evans's own words.

Bath, April 23, 1787.

Mr. Evans is since dead; but the disorder which carried him off, had, I am informed, no connexion with any complaint of the urinary passages.

CASE XIV.

The following Case is that of M. S. Branthwayt, Esq. of Taverham near Norwich, as related by himself.

" In June 1785, I was seized with a slight fit of the gout; and during the time the gout was upon me, I felt a violent pain quite round my body, but more particularly in my right kidney, attended with a frequent inclination to make water, which passed from me very slowly, and with much pain. At this time Mr. B. Colborne, of Bath, came to visit me, and upon inspecting my chamber-pot, found that my water was coffee-coloured, and, from my description of the symptoms I had felt, suggested to me that I had the stone. Fortunately Mr. Benj. Colborne had a small travelling apparatus with him to make the Aqua Mephitica Alkalina: he made me some, and after I had taken 4 or 5 doses, my urine passed more freely, and with less irritation, and became perfectly clear. When he left me, I was without the Aqua Mephitica Alkalina three weeks, in which time I voided a small stone, and some red gravel, and my urine became turbid, and gave me infinite pain in passing. In the space of two or three days after, I began again to take the Aqua Mephitica Alkalina; my water became clear, and passed without irritation, and I continued entirely free from all symptoms of the gravel, and was in better health than I had been for some time before. Indeed, feeling so perfectly well, I left off the water for 5 or 6 months, when I was again attacked with a violent pain in my right kidney, attended with fever and sickness, which lasted two or three days; but not suspecting it was the stone, and finding myself very well again, I did not take the Aqua Mephitica Alkalina till after I was a second time attacked in the same way, but more violently. I then took the water again, and was perfectly well till September 1786, when I voided a stone about the size of a vetch; and since that time I have never left off the Aqua Mephitica Alkalina, nor have I

had

had any symptoms of the gravel or pain in my kidney.

Taverham, near Norwich.   M. S. BRANTHWAYT.

"Finding so much benefit from the Aqua Mephitica Alkalina, I recommended it to a poor woman in my parish (by name Kidd), who has been afflicted with the stone and gravel at different times between 20 and 30 years, during which time she has voided many small stones. This poor woman being always an invalid, passing by her door, I called to ask her how she did, and found her very ill with a nephritic complaint. I made her some Mephitic Alkaline Water, and after taking a few doses she was much relieved, and continued mending for three weeks, when she voided two stones, one near an inch long, the other about the size of a pea, weighing together 24 grains. She continued drinking the water some time after, during which time she had no return of her gravelly complaints, and thought herself in better health than she had been in for many years. I tried to persuade her to continue the Aqua Mephitica Alkalina; but she said she had no symptoms of gravel, and was very well; therefore I could not succeed.

M. S. B."

Extract of a Letter from M. S. Branthwayt, *Esq. dated* December 14, 1788, *from* Taverham, *near* Norwich, *to* Benjamin Colborne, *Esq.*

" *In answer to your letter respecting the Mephitic Water, I must inform you, that I have very constantly drank a large glass every morning, when at home, before breakfast. I have been perfectly well in health except gout, and have not had the least return whatever of my gravelly complaints.*"

December 5, 1791.

I was this day informed by Mr. Colborne that he had received a letter from Mr. Branthwayt, dated November 30, 1791, *in which he says, that he has not taken any of the Alkaline Water for ten months, and has not felt any symptoms of gravel. He has, however, had a fit of the gout more violent than he had before experienced.*

CASE

CASE XV.

Adrian Abbot, cabinet-maker, being in London in the year 1775, and employed in his bufinefs, whilft he was carrying a coffin, it being high above his head, it ftruck againft a building, and ftrained him in the fmall of his back. Three days after this accident he paffed a confiderable quantity of blood as he went to make water. From that time to the year 1783 he had frequent inclinations to pafs his water, and a conftant pain in the fmall of the back, and frequently a mucous difcharge of a light yellowifh colour, if he exerted his ftrength to any confiderable degree. Some time afterwards he was feized with a total ftoppage of his urine, for which he was bled by the advice of a furgeon, and took medicines, and in about a week his urine paffed eafily. About fix months afterwards he had another attack of the fame kind, which, when it abated, was attended with a large difcharge of blood, which became from that time a frequent fymptom. At this time it was the opinion of two furgeons who attended him, that he had a ftone in his bladder.

der. He continued in this ſtate, with great pain in his back, and frequent inclinations to paſs his urine, until December 14, 1786, when, being worſe, he applied to a ſurgeon, who directed him ſome medicines, but without ſucceſs. He then applied to another gentleman of the profeſſion, who adviſed him to ſubmit to be cut. His urine at this time was foul and turbid, and changed the vegetable blues to a red colour.

On the 27th of January he was directed to Mr. Benjamin Colborne, who gave him ſome of the Mephitic Alkaline Water, with directions how to take it. His urine ſoon became clear and tranſparent, and he could retain it a long time, and for many nights had occaſion to make water once only. The pain in his back ceaſed, and at preſent (April 20) he has no complaint of any kind.

This perſon is, ſince the publication of the above caſe, removed from Bath to Briſtol; but Mr. Colborne has very lately received information, that he is quite well, and has had no return of his complaint, but ſtill continues taking the Mephitic Alkaline Water.

CASE

CASE XVI.

A tradesman of this city, of the name of Ralph, had for many years been troubled with a stricture, occasionally, of the urethra, which caused frequent obstructions to the passage of his urine, for which he was obliged to use bougies, which gave him great pain without producing any permanent relief. During the spring of the year 1785, his complaints returned so frequently, and with such violence, that it was not practicable to pass a bougie. In this melancholy condition nothing but opiates afforded him any relief, which were obliged to be large in quantity, and frequently repeated. In the month of May in the same year, he had an abscess in perinæo formed, which discharged a large quantity of matter, and healed in about a month's time. His difficulty however of passing his water continued, and his urine was loaded with mucus. Nearly in this state he continued until the eleventh of December 1786, when he began to take the Mephitic Alkaline Water.

At that time his urine came from him drop by drop, and was so overcharged with mucus that it adhered to the pot in such a manner, as not to fall out when the vessel was turned with its mouth upwards. Its smell was also so rank and fœtid, as scarcely to be borne, and was so alkaline with the putrefaction, as to ferment with oil of vitriol, and to change paper stained with juice of turnsole to a blue colour. His pain was likewise so great as to oblige him to take opiate pills every three or four hours. He began the use of the Mephitic Alkaline Water by taking it with milk as before directed, and with some drops of laudanum in each dose. He took it every eight hours, and in the space of twenty-four hours he passed his water much easier, and his irritations were less frequent and violent, and his urine came away in larger quantities at a time. By continuing his medicine he was able, in three or four days time, to take it without the laudanum, and to go about his business, which is that of a shoemaker, and to work at it much better than he had been able to do for several years. Since the above-mentioned time, he has had little or no return of his complaints, and is in every respect better in health than he has been for the last

seven

seven years. Before he took the Alkaline Mephitic Water, he was not able, even when he thought himself tolerably well, to work at his bufineſs for two days together. Although this man has taken the Alkaline Mephitic Water for four months, it seemed to have very little effect on a fragment of the same calculus, mentioned in the experiments before related to be made with urine. Yet it appeared to be of an alkaline nature, by its changing paper ſtained with juice of turnſole to a blue colour.

Mr. Ralph continues perfectly well, but ſtill uſes the Mephitic Alkaline Water, which is now become very agreeable to his palate. He grows fat under the uſe of it, and his complexion becomes florid.

December 1, 1791.

Mr. *Ralph is now quite well, and has not taken any of the Alkaline Water for the ſpace of ten or twelve months.*

CASE

CASE XVII.

―――― Telling, a glazier of this city, had been for two or three years subject to strictures of the urethra, which produced frequent suppressions of urine. About ten months ago from the present time, May 1, 1787, he was taken ill in the last-mentioned way, and continued without passing any water for two days, in which he suffered the most excruciating pain. A bougie was passed, which produced a discharge of urine for the time. He had frequent returns of pain of this kind, but not so violent, and was never easy two days together. When he went to make water he had in general a tenesmus, and in the morning his urethra used to be greatly clogged by viscid mucus, which delayed his urine passing, and frequently produced a temporary stoppage. His urine was foul in appearance, and of a wheyish colour, with gravel in it, and made in small quantities.

After taking the Alkaline Mephitic Water for three or four days, his urine began to pass more freely and easily, and his complaints mended

mended daily. He has now taken the Mephitic Alkaline Water for two months, has no tenefmus on paffing his water, can retain it well, and pafs it in large quantity, and the mucus is but little, and that of a thin confiftence compared to what he formerly voided. He eats, drinks, and fleeps well; is able to labour in his bufinefs, and is upon the whole in better health than he has been for feveral years.

This perfon continues perfectly well, although he has left off taking the Mephitic Alkaline Water.

December 1, 1791.

This perfon continues quite well, and has left off taking the Alkaline Water near two years.

CASE XVIII.

The Reverend Henry Wilfon, Vicar of Heverfham, near Kendal in Weftmoreland, aged 69 years, had been for fome time afflicted, principally at intervals, with a complaint of the urinary paffages. This was much aggravated by wet or cold, and even by moderate exercife exercise or motion. He could not walk a quarter of a mile without passing coffee-coloured water, and frequently little besides blood. This was accompanied sometimes with great pain, and always with much uneasiness. His water encrusted the chamber-pot, and deposited besides a quantity of loose red sand. Sometimes filaments resembling bits of skin, might be seen floating in it. His appetite was but indifferent. In this state he began, on the 10th of May, 1788, to drink the Mephitic Alkaline Water. Of this he took at first a pint and half daily, divided into three doses; one of which he took an hour before breakfast, a second at eleven o'clock in the forenoon, and a third at six in the evening. This he soon afterwards reduced to a pint daily, taken in divided doses at the periods above mentioned. He took it at first with hot milk, as directed in this work: this he changed for raison wine, and this again for ginger wine, but again returned to milk. He pursued this course for three months, with some, though no great alleviation of his symptoms. His appetite was amended, and his urine left no crust on the chamber-pot. The above was the state of this gentleman's health, as described by him

him in a letter I received from him in the beginning of September laſt. His laſt account, however, is much more ſatisfactory.

In a letter dated November 7th, 1788, he ſays, " Every diſagreeable circumſtance attendant on my complaint, has now left me. I find my appetite greatly improved, my health extremely good. I am as equal to any exercife of walking, &c. as I have been for ſeveral years, though entered upon my 70th year.

" Cold and wet were ever inimical to my diſorder; Sunday duty, therefore, or a funeral, or being a few minutes in a ſhower, brought on my complaint, which continued troubleſome for three or four days before it left me: but I feel no inconvenience at preſent by being confined in my church for more than an hour and half at once; nor in being expoſed to a ſtorm of wind and rain for the ſame ſpace of time.

" I muſt own it requires a ſtrong reſolution, and no little faith in the efficacy of this medicine, to continue the uſe of it long together; and I drank it for more than three months, before I perceived any benefit from it;

it; but it was extremely cold to my ſtomach moſt of this time, and diſagreeable; nor would I attempt to warm it with brandy, or any ſpirits, all the while. However, the Biſhop of Llandaff, and yourſelf, bad me perſevere. I did perſevere, and am it this time enjoying the fruits of this obedience; and I could not excuſe myſelf, did not I here offer you my ſincereſt thanks for the bleſſing of the freedom from pain, &c. which I now enjoy."

Extract from a Letter, dated December 24, 1788.

" I have for ſome time paſt been returning viſits on foot, at the diſtance of two or three miles, and twice or thrice a week. On Friday ſe'nnight I was celebrating the birthday of a lady in this neighbourhood, from whence I returned about three o'clock in the morning of Saturday, without being in the leaſt diſordered with ſitting up, or my morning walk.

" I can with the greateſt truth and pleaſure aſſure you, that my health and appetite are both of them now, and have been for ſome months paſt, remarkably good; for
which,

which, under God, I think myfelf indebted to the Mephitic Alkaline Water.

Extract of a Letter from the Rev. Henry Wilfon, *dated* October 22, 1791.

"*With the greatest truth and pleasure, dear Sir, I now assure you, that, since my last to you, near three years ago, I have, thank God, remained entirely free from calculous complaints, my health perfectly good*—cruda mihi viridifque fenecta. *Yet I have continued the use of the Aqua Mephitica Alkalina regularly, about half a pint at eleven o'clock in the morning every day. However, about a year ago, I was prevailed upon to omit for a month the use of the Mephitic Alkaline Water; but I suffered, alas, for my credulity! All my former complaints returned, seemingly with redoubled pain; yet, by resuming immediately the use of the water, I was quite free from all my complaints in two or three days. At this time I passed a stone of the size of a common apple-pippin, and of a light brick colour; it had lain in the pot some hours, and had lost part of its weight and size. If you, Sir, can make the above of use in this your fourth edition, it is very much at your service.*"

CASE XIX.

" The Rev. Robert Burton, of Oakingham in Berkshire, aged 67 years, of a robust habit, and temperate in his way of living, though subject to the gout, had been used to pass red gravel in his urine, and sometimes small calculi, about the size of a large pin's head. About eight years ago, he had a violent attack of the bleeding piles, to which he had been before subject. About this time his water often came away resembling coffee-grounds, and, if he took any exercise, appeared to be no other than pure blood, but without much pain. These bleedings, however, were moderated by a course of the bark. He had in the beginning of August (1788) a fit of the gout, which he had not had to any great degree for ten years before.

" After this, his pains in making water were very great; the calls frequent, and the quantity small, attended with shiverings, and an exceedingly troublesome irritation backwards, except when he was lying down in bed. At this time he was supposed to have a ſtone in his bladder, and was founded, but no calculus was diſcovered. He paſſed every morning a large quantity of viſcid mucus, of a dark colour, and ſometimes mixed with a little grumous blood, but not more fœtid than common urine. The acid and alkaline draughts (as recommended by Dr. Hulme) were then tried; but as no effect was obſerved from their uſe in the ſpace of three weeks, they were laid aſide, and lime-water made from oyſter-ſhells ſubſtituted in their room. The patient, however, growing worſe, the lime-water was left off, and an infuſion of the red bark adviſed and perſiſted in until a ſevere flux obliged him to lay aſide all medicines except aſtringents and opiates. When this ceaſed, he returned again to the bark, which he took thrice a day, with five drops of laudanum in the laſt doſe. He had, at this time, a continual thirſt, though no fever, and almoſt every other ſymptom incident to ſuch complaints, to a very diſtreſſing degree."

The above ſtate of the caſe was ſent to me, dated September 23, 1788, deſiring my opinion. I adviſed a continuation of the uſe of the bark in ſmaller quantity, and to try the alkaline water with hot milk, as directed above,

above, to the quantity of half a pint daily, taken at two dofes, and with a few drops of laudanum, and a little manna to be occafionally taken if coftive.

In a Letter, dated the 22d of November (1788), I received an account of the effects of the remedy as follows:

" Of three dozen of Killick's half pints, I have taken twenty-one in hot milk, with laudanum, as well as the red bark, as directed; only the laft proving too aftringent, and the means to counteract it throwing me into the contrary extreme, I left it off. Whilft coftive, I had feveral bloody ftools, though no bloody urine as ufual with me in that ftate; and to be lax, is always diftreffing, and heightens my fufferings: fo I now take the medicated water only.

" As to my prefent ftate, my appetite is good, and my fleep comfortable, and as found as may be with fo many interruptions from urinary calls, which, though much lefs frequent than they were a fortnight ago, are frequent enough to require the urinal in bed; the ufe of which obliges me to lie nearly on

my

my back. In this pofture, my water paffes fo eafily, that, on waking, I often find the veffel quite full. At other times, the paffage of my urine requires efforts, but now feldom occafions any troublefome irritations backwards. For about a week paft I have moderately fucceeded in keeping my body regular; my urinary calls have not been near fo frequent as formerly; and there has been fome diminution in the quantity of mucus, though enough of it yet paffes every night to fhew that the caufe of the complaint ftill exifts. Whatever this is, it feems to be feated about the neck of my bladder; for there the feeling is at one time like what one would expect from a fore expofed to the paffing urine; and at another, a kind of itching like that of a healing wound. I have watched attentively to find fome fragments or gravel, but in vain. My general health is far better than it was three weeks ago, though the gout or rheumatifm, for it is not yet fwelled, has fallen upon my knee. Such is my prefent ftate."

I have fince received another Letter from Mr. Burton, dated December 15, 1788, in which he fays, that about a fortnight after he wrote the letter of November 22d, he paffed a thin gritty fubftance, of the fize of a filver penny, which he thinks to be a fragment of a calculus. At prefent (December 15, 1788), he fays, that his urine is very clear, and that he retains it fufficiently to make the intervals between his calls about an hour and three quarters in length: his fleep is comfortable; his appetite and digeftion good; and his health in general (as he thinks) in a fair way of being re-eftaplifhed; and himfelf reftored, as he expreffes himfelf, to a capacity of enjoying a good fhare of the comforts of life.

CASE XX.

James Clayton, Efq. of Cavendifh Square, London, was firft attacked with the gravel about the latter end of the year 1785, and had fits of it once or twice a week, attended with great and frequent irritations of the urinary paffages, and often brought away fmall ftones about the fize of pins heads, and fome much larger; of different fhapes and colours; fome of them being yellowifh, and others white; all which were attended with great pain when they came away. He tried many medicines for the above complaint, but without

without much effect. He drank marsh-mallow tea with the seeds of the wild carrot, and barley water with gum arabic, but without benefit.

On the 30th of July, in the year 1787, he began to drink the Mephitic Alkaline Water, of which he took seven ounces daily. In the first three months after he began to take it, he had but six or seven attacks, which was not above half the number he had before in the same space of time. About the middle of October 1787, he had several attacks in the space of two days; and in one of these, which was very severe, he voided five or six calculi at once; and in the space of six hours, seven or eight more; some of them as large as, and resembling in shape, half cherry-stones, and others rough and triangular; but most of them concave or convex, as if they came from a stone as large as a small walnut. During his pain, he was frequently obliged to have recourse to emollient clysters, and sometimes to laudanum, which he took from 30 to 60 drops when in extreme pain; but was desirous to avoid it, as it injured his appetite, which, if the laudanum was not taken, appeared to be amended by the Mephitic Alkaline Water,
which

which agreed alſo well with his health in every reſpect. In the beginning of November 1787, he diſcharged ſeven more calculi, larger than any he had brought away before, and of a whitiſh colour, and convex and concave in their ſhape, and as large as cherry-ſtones. Theſe, as might be expected, produced great torture in their paſſage, and great ſoreneſs of the parts for ſome time after, and a ſenſation of weight at the neck of the bladder. One day towards the middle of the month (November 1787), he was ſeized with great pain in the right kidney; and in an hour or two found a calculus coming forwards, which about three in the morning came away, and proved to be the largeſt he had ever paſſed. In half an hour's time after the firſt came away, he paſſed ſeven others of a ſmaller ſize. Theſe were followed with ſuch pain in making water as was even greater than when the ſtones were paſſing. His urine was cloudy and mucous, and depoſited a red ſediment on the pots: about this time he was accuſtomed to uſe occoſionally a warm bath, and a fomentation of camomile flowers in a flannel bag, which gave temporary eaſe. Before he took the Mephitic Alkaline Water, the ſtones he paſſed were yellowiſh and hard; but ſince he drank it, they

they have become smaller in size, white, and crumble on pressure between the fingers. In January 1788, he passed two smooth stones, one as large as a vetch, and the other the size of a barley-corn, without much pain. Since January 8th, 1788, to this present time (December 2d, 1788), he has not had above twice a little irritation in passing his water, which did not last an hour each time; and since February last, has been perfectly well. He is now able to travel in his chariot for twenty miles together without pain, and to walk a mile and half; whereas, before he took the Mephitic Alkaline Water, he could not walk across the room. He has taken the quantity of half a pint daily, with about half a tea-cup of warm milk in each half pint of the Water, since January last (1788).

The above account is literally true.

Cavendish-Square,
Dec. 2, 1788.
JAMES CLAYTON.

December 1st, 1791.

Mr. Clayton is since dead, but of a different complaint from that which was the subject of the foregoing Letter.

CASE

CASE XXI.

Copy of a Letter from Lieutenant-Colonel Williamſon, of the Royal Regiment of Artillery, to a friend, on the ſubject of the Stone and Gravel, with which he was afflicted, and the ſurpriſing benefits derived from the uſe of the Aqua Mephitica Alkalina.

Shooter's Hill, Oct. 15, 1788.

" Dear Sir,

" It is with great readineſs that I ſit down to give you an account of my complaint (the ſtone and gravel), as alſo the important ſervices which I have received from the uſe of the Aqua Mephitica Alkalina.

" In the latter end of the year 1781, when at Barbadoes, I was ſeized with a violent ſtrangury, and pain in my kidneys, which confined me to my bed near a week: fomentations, and warm-bathing, afforded me temporary relief; but from that period I was ſcarcely a fortnight together free from pain. In July, 1782, I had another moſt violent attack, attended with

an

AQUA MEPHITICA ALKALINA. 93

an acute pain in my kidneys, a numbness in my thighs, with great sickness at my stomach, and an head-ach; here my former treatment failed, and the medical gentlemen advised me to embark for England, which I did; my disorder continued with little intermission from pain to the 24th of September, when on my passage home, and after six days torture, there came from me a stone about the size and shape of a lemon-feed: from this time until July 1787, I was never ten weeks free from pain, and passed above one hundred and fifty stones (exclusive of small gravel); some larger than the first which I passed, and few smaller than an hemp-feed. I had the best medical advice whilst in England, and adhered strictly to their directions for some years; soap-lees, four and five pounds of honey in a week, wild carrot, and flax-feed tea, and numberless other experiments I tried, but without relief; bloody urine, loss of appetite, and continual pain, had so much debilitated me, that I could scarcely walk, stand, or ride; and I almost began to despair ever of meeting with any thing that could give me ease, or remove my complaint: but early in June 1787, fortunately for me, a gentleman who resides mostly at Bath did me the honour of a visit, and strongly recommended

Mr.

Mr. Colborne's Alkaline Solution, impregnated with Fixed Air, from which I have derived the happiest effects. It is now more than sixteen months since I have adopted the above preparation, during which time I have been totally free from my disorder; my appetite is returned; I can take my exercise as usual, and I indulge myself in several favourite things in eating and drinking, from which I was before debarred, by advice of physicians. I allow the solution to be exposed to the Fixed Air for sixty hours, instead of forty-eight, (as mentioned in the printed instructions delivered at Parker's glass-shop, in Fleet-street,) which agrees better with my stomach; and I restrict myself to half a pint of the Water on rising in the morning, and the same quantity on going to bed. I hope that the above statement of the effects of the Alkaline Solution, impregnated with Fixed Air, upon my disorder, may induce your friend to make an experiment of it; and that he may receive the same benefit with myself from it, is the sincere wish of,

   Dear Sir,
    Your most faithful,
     humble servant,
     JOHN WILLIAMSON."

Letter from Colonel Williamfon, *to* Benjamin Colborne, Efq.

Shooter's-hill, Nov. 29*th,* 1791.
" *Sir,*

" *I am happy to have it in my power myfelf to acknowledge your Letter addreffed to Mrs. Williamfon; and ftill more happy, that the account which I can give of my health is fuch as is extremely gratifying to myfelf, and I truft, from the intereft you take in thofe who have been benefited by your exertions, will prove highly fatisfactory to you. In June* 1787 *I firft attended to the Mephitic Water. In my ftatement which appeared in the laft edition of the Virtues of the Aqua Mephitica Alkalina, I expreffed the advantages I had derived from it: fince that period I have been perfectly free from even a diftant fymptom of my former diforder, with only one exception. In a long paffage of feventeen weeks* (1790) *from the Weft Indies, I was under the neceffity of being an economift of the Water, which I had bottled, and difcontinued my ufual practice of half a pint of it in the morning, for about three weeks: nearly at the expiration of that time, I was fenfible of fome alarming fymptoms;*
I inftantly

I inftantly had recourfe to the few bottles I had in referve, which difpelled every uneafy fenfation; and landing fhortly after, I had an opportunity of recruiting my ftock. I am now, and have been ever fince my arrival in England, perfectly well; ftill adhering to the Water, but taking half a pint only in the morning. Permit me to fubfcribe myfelf, what I really am,

Dear Sir,
Your very obedient
and obliged fervant,
JOHN WILLIAMSON."

C A S E   XXII.

Communicated by Mr. Perry.

A lady of this city, who wifhed her name might not be mentioned, aged fifty, of a thin habit of body, for many months has been afflicted with great naufea, lofs of appetite, violent pains about the region of the kidneys, frequent and forcing pains to make water, which comes away in very fmall quantities at a time, and, on fettling a few minutes, depofits a vifcid mucus, and fand. A great variety of mucilaginous and oily medicines have been

taken

been taken to no effect. On being informed of her symptoms, I recommended two ounces only of the Alkaline solution, impregnated with Fixed Air, to be taken three times a day. It agreed perfectly well with her stomach: in the course of a week she found herself sensibly better in her general health; the urine became clear, and the remaining nephritic symptoms left her. The lady has continued the solution one month, and is exceedingly well without taking any other medicine.

CASE XXIII.
Communicated by Mr. Perry.

Thomas Shell, of this town, aged 13 years in September 1787, applied to me, at the request of Mr. Colborne, to be sounded for the stone. He had great pain and difficulty in making water, which came away by drops. He also complained of a bearing weight at his fundament, where nothing uncommon was to be perceived. From his symptoms, I sounded him, and found a stone in his bladder. I desired him to inform Mr. Colborne of this circumstance; which he did, as I was informed, by the gentleman, who humanely and generously took him under his care, and administered the solution with more than usual good effect.

fect. I believe the medicine was used for twelve months, at times. Within these ten days the lad called on me, with the greatest pleasure, to found him again, which I did repeatedly, and could not perceive any stone, nor did he complain of the least symptom of it, December 4, 1788.

He took the water two months before he found any benefit.

December 1, 1791.
This patient has not taken any of the Mephitic Alkaline Water since December 1788. *He is now in perfect health, and has been so ever since the time before mentioned.*

CASE XXIV.

John Fussell, of Bath, about 13 years old, who had been cut for the stone in Bristol Infirmary, six years before, could never retain his urine in the day time, from the time of his being cut. He began taking the Mephitic Alkaline Water November the 29th, 1787, and drank in two days a quart bottle of it, with 30 drops of laudanum and some hot milk. This was continued for about a month, and then

then the laudanum and milk was omitted. He had not taken the water above a month or six weeks, before he retained his urine perfectly well; which he continued to do, whilst he took the water; but on leaving off the use of the water for a few days, his disorder returned, and continued as before mentioned, for a month, till he began again taking the water, which had the same good effect as it had before for three or four months. But by leaving off the use of the water a second time, his disorder returned as before. About the beginning of October 1788, he began again on the water, and in five or six days time he could retain his urine again, and did so till December 2, 1788; and then, by his own neglect in not continuing to take the water, though but for four or five days, his disorder returned a third time; but now he promises not to omit it for the future, and is likely to do well.

December 1st, 1791.

Whilst John Fussell was in a course of taking the Mephitic Alkaline Water he retained his urine perfectly well; but he has not taken any for seven or eight months. He now says, that whilst he continues at his work in a sitting posture he can retain his urine; but that, if he uses much walking exercise, it comes from him in small quantities.

CASE XXV.

Copy of a Letter from Dr. Bourne, *Phyſician at Oxford, to* William Falconer, M. D.

DEAR SIR,   Oxford, *May,* 6th 1790.

" I read your " Account of the efficacy of the Aqua Mephitica Alkalina in calculous diſorders, &c." with much ſatisfaction; and having met with an inſtance of its good effects, I think it proper to ſend the caſe to you, that you may make what uſe you pleaſe of it in a future edition.

" Mr. Goſwell, the ſubject of the following caſe, is a reſpectable, plain, ſenſible man. His good underſtanding enables him to deſcribe his ſymptoms clearly, while his want of ſcience is a ſecurity that he does not bend circumſtances to any medical notions of his own. With regard to myſelf, I cannot be ſuſpected of exaggeration, as it will be ſeen that I was entirely unconcerned in directing the means which relieved the patient.

    I am, dear Sir,
        Your obliged
           and faithful ſervant
                ROBERT BOURNE.

January 23, 1790.

"Mr. William Gofwell, dealer in timber, of Woodftock in Oxfordfhire, now fifty years of age, had, previous to the year 1786, been feveral times afflicted with pain in the neighbourhood of the kidneys, which was fometimes attended with difficulty in making water; twice, when thus affected, the pain was fo confiderable as to induce him to fubmit to the exhibition of a clyfter, which immediately relieved him; and he formed no other opinion of his complaint than that it was cholic. In the autumn of the year 1786, he was fuddenly feized with a violent pain in the right kidney, extending from thence acrofs the lower part of the abdomen towards the bladder: the pain brought on vomiting, and lafted forty-eight hours, without intermiffion; the water, during this time, came away in drops, and was bloody: at the end of forty-eight hours he became eafy, the urine then paffed freely, and was no longer bloody; he continued eafy for feven or eight days, when, riding on horfeback, he had a fudden call to make water, but found, when he attempted to obey this call, that he could void fcarcely any, and the attempt brought on confiderable pain in the urethra.

He returned home, and drank an infusion of the garden parsley, refraining as long as he could from any effort to make water, thinking that the fuller the bladder became, the more force he should be able to exert against the obstructing cause; by and by, on straining hard, he forced away a stone from the urethra, shaped like an orange-seed, and nearly half an inch in length.

" From this time to the summer of 1787, he continued easy, and made water freely; at the time last mentioned he was again suddenly seized with a pain in the right kidney; this lasted about an hour, and then went off: he kept still for many hours; the pain did not return while he kept still, and he made water with ease; but, on attempting to move about as usual, the pain in the kidney returned, though with less violence. He had frequent calls to make water, made but little at a time, and that of a coffee colour, from which blood subsided on standing. From this period exercise on foot or on horseback constantly brought on the painful sensation in the right kidney, a pain extending from the hip down to the knee on the right side, a sensation of weight across the upper part of the ossa pubis,

a frequent

a frequent inclination to make water, which was always on thefe occafions voided in fmall quantities at a time, and exhibited the appearances above mentioned: he was eafy no longer than while he kept himfelf in a ftate of reft.

" In this fituation, a very diftreffing one to a man of an active mind and in an active bufinefs, he remained more than a year and half; in which fpace he had good medical affiftance, gave a fair trial to many medicines, and obferved great regularity in diet. Among the medicines were lime-water and the infufion of the wild carrot feeds. The limewater- induced an unpleafant ftate of coftivenefs; but he did not think that it at all relieved his complaint: of the infufion of the wild carrot feeds he fpeaks more favourably, and is clearly of opinion that at times it gave him confiderable relief. He obferved that during this period, his chamber-pot was always covered with a pretty thick fur, and that he fometimes voided bits of gravel.

" About April 1789, when his complaint had rendered him more thin, weak, and difpirited, Mr. Knipe, a clergyman who ferved a church in his neighbourhood, became acquainted with his case; and recollecting that a friend of his, in a similar situation had been re...ved by the Mephitic Alkaline Water, he .. .. .. your treatise to Mr. Got...., who imm............ procured a machine for pr..p..... the m........ine, prepared it according to yo.r ..rections, ..nd took the third of a pint, th...c ..mes a ..ay, without any addition: at the end of .. little more than a month he began to feel himself benefited, and in two months found himself manifestly better; from that time he ventured gradually to increase his exercise, and experienced no inconvenience from it; he proceeded cautiously, and was soon able to attend to his business, and rode or walked after it, as suited his convenience: latterly he has sometimes been upon his legs five or six hours in a day, or has ridden eight or ten miles and back again; more than once he has ridden forty miles in a day, in a stage-coach; and none of these exertions have caused a return of his complaint.

" Since his amendment the chamber-pot has not been furred in general, nor has he voided bits of gravel of any size; but he has sometimes voided a little fine sand. He can now lie all night

night without ufing his chamber-pot; whereas before his amendment, even when eafy, he was obliged to ufe it four or five times in the courfe of a night. He cannot help thinking that the quantity of urine has been greater fince the taking of this medicine than it was before his indifpofition, though the quantity of liquids, which he drinks is not greater. He was difpofed to coftivenefs before he took this medicine; that difpofition is now removed, and the body is kept regularly open by it. He has recovered his flefh and fpirits; his appetite is good; but he fays, that did not fail him much during his illnefs.

" He took a pint of the water daily for fix months; fince that time he has taken two thirds of a pint only."

Copy of a Letter from Mr. William Gofwell, *to* Dr. Bourne, *Oxford.*

SIR, *Woodcott, Nov.* 28, 1791.

" Since I had the honour of explaining my cafe to you in January 1790, have had no return of my old complaint. I have fometimes obferved fome fandy fediment at the bottom of the chamber-pot, but felt no pain or uneafinefs easiness usually attending the gravel and stone, which I had so long before, and so severely felt; but, thanks be to God, have enjoyed exceeding good health were since; and am,

  SIR,
   With all due respect,
   your most humble servant,
  WILLIAM GOSWELL.

" P.S. I still continue taking the water, nearly half a wine-pint each day: I generally take it fasting, and last at night."

C A S E   XXVI.

Extract of a Letter from Mr. Samuel Bentley, *of Uttoxeter, Staffordshire, to* William Falconer, *M. D.*

DEAR SIR,   *Uttoxeter, Aug.* 12, 1790.

" As I have received such benefit from your publication, which directs the way of making the Aqua Mephitica Alkalina, and as I feel myself under so great obligations to you; I think I am bound, out of gratitude and justice, to send you my case, with an account of the effect that medicine had upon me; which
             I think

I think will appear to the world to be as wonderful, and as ſtrong a proof of the efficacy of the medicine, as any you have yet publiſhed.

" It will be proper, in the firſt place, to let you know, that from my early youth I have always been of a tender and thin habit of body, ſtill rendered more ſo, by having the ſmall-pox in a very bad way; but though I am now upwards of ſixty, I have, till the laſt ſixteen or ſeventeen years (except being frequently troubled with the piles) had tolerable health, ſo as to be able to follow my buſineſs, which was not indeed of a kind that required any violent exertions, except riding pretty long journies; and as I had ſufficient leiſure from my buſineſs, I had alſo ſpirits to enjoy ſeveral amuſements, particularly bowling in the ſummer, and going out with my greyhounds in winter, and could follow them moſt part of the day without fatigue. And I muſt alſo add, that I was very careful, not to indulge in any exceſs in eating, and more particularly in drinking.

" About the time I mentioned above, I had frequent rheumatic complaints, and my health began to decline greatly; I was often much

out

out of order, had frequent shiverings attended with feverish and hectic disorders, with pains about my loins, and often so very weak and low, that I fell into fainting fits after using the least exercise; so that I was obliged to decline both my business and several of my amusements. I frequently found quantities of small red sand in my chamber-pot, but did not then think that the gravel was my principal complaint; though I afterwards found, that a calculous disorder grew upon me every year: the gravel that came from me grew larger, and I often parted with stones about the size of a vetch; and whenever I got a cold, the calculous complaint came upon me with greater violence.

" I continued much in the same way till the spring of the last year, 1789, when I grew so bad that nothing which had used to relieve me gave me the least ease. I had the advice of the medical gentlemen in the place where I live, and all the usual remedies were tried, but without success; and though I followed their prescriptions with the greatest exactness, my pains continued with the same violence, attended with all the aggravating appendages to that terrible complaint; I was not able to

ride

ride out a little way, though I went the moft gentle pace poffible; and I could not walk in the garden without my water being moft part of it blood, and it would even be fo if I did but walk about in the parlour; I had befides fuch frequent urgings and irritations to make water, though it was only for a few drops, that they came upon me ten or twelve times in a quarter of an hour; and the acrimony was fo great, that after the laft drop the pain was almoft death to me. I was often obliged to change my fhirt for a dry one, and that would be in the fame wet condition in five minutes time; fo that for fome weeks I was even offenfive to myfelf; and at laft I was as bad as ever any perfon could be, and my pains as intolerable; and as calculous diforders have been hereditary in my family, and proved fatal to fome of my anceftors, I began to give myfelf up as incurable: when (happily for me) the lady of the nobleman who franks this letter, hearing of the deplorable way I was in, fent me your book of cafes, with the directions for making the Aqua Mephitica Alkalina; and as I faw fome cafes in it nearly fimilar to my own, particularly Dr. Cooper's, I immediately fent for a glafs apparatus from Parker; and as I had, among other

other acquirements, some knowledge in chemistry, I soon got into a way of making the medicine, so as to be perfectly saturated with the Fixed Air; and I am happy to inform you, that after I had taken the medicine twice a day for about a fortnight, I began to find benefit: the first appearance of amendment was in my water, which began to get clear, and more free from mucus; and after some time I made no more bloody water: I could however still perceive I had stones either in the neck of the bladder, or the urethra; but though they continued to feel uneasy, the sting of them was gone. The first time I ventured to go out, after I was better, was about five miles in a post chaise, to return thanks to the lady who sent me your book of cases; and the day after, I parted with a stone, about the size of a large pea, one side of it much corroded, and it came from me without the least pain: I parted with several more afterwards still more corroded, so that they crumbled betwixt my fingers; which, I think, proves the efficacy of the medicine beyond a doubt: but the last stone I parted with, which was about three months after I began with the medicine, had all the inside entirely wasted away, being nothing more

than

than a ftony cave, interfected with fine fibres in every direction, like a cobweb; but in attempting to wrap it up in paper, in order to preferve it, I broke it to powder.

" I began to take the medicine about the 18th of July laft year, and I took a quarter of a pint tumbler of it twice a day till the 1ft of January; and fince that time I have taken it only once a day, which I find keeps me perfectly free from pain, as alfo from any fymptoms of my former fufferings: I have no more thofe urgings, irritations, and acrimony, and I can now hold my water from three to four or five hours; and I can now take my morning walks into the fields as ufual before breakfaft, and my rides for ten or twelve miles betwixt breakfaft and dinner; and I do not think I have any particles of calculi remaining; and the medicine not only gives me fpirits, as much as if I drank a glafs of Champagne, but agrees with me fo perfectly well in every refpect that I have recovered my flefh again, fo much fo, that I have been obliged to have all my cloaths let out.

" I fhould have wrote to you fooner, to have informed you of the benefit I have happily
received

received from the medicine, but I waited till I had given it one whole year's probation, which is now more than completed. I am rather in doubt whether I ſhould continue ſo well, if I was to leave it off entirely; ſo I take a little tumbler of it once a day, and ſhall do ſo for ſome time longer, as it agrees with me ſo perfectly well.

"Pleaſe to accept of my moſt grateful acknowledgments, and my ſincereſt thanks for the infinite ſervice your remedy of the Aqua Mephitica Alkalina has been of to me; and I am, with the trueſt eſteem,
Dear Sir,
Your highly obliged
And very humble ſervant,
S. BENTLEY."

Extract of a ſecond Letter from Mr. Bentley.

"SIR,   *Uttoxeter, 24th Nov.* 1791.

"I think myſelf honoured by your favour of the 13th inſt. I was anxious to get a frank, that I might take the firſt opportunity of anſwering it, which I now do with the utmoſt gratitude to you for being the means (through the goodneſs of Providence) of reſtoring me

to

to that greateſt of all bleſſings health, from a ſtate of ſuch deplorable miſery, which none can conceive, who have not been in the ſame ſad ſituation; and I muſt alſo inform you, that the benefit I received from the Alkaline Water has proved permanent, with reſpect to my calculous complaint, and ſo every way beneficial to my conſtitution, that it has relieved me from the faintings I was ſo ſubject to, hectic heats, &c. And I now write to you in joy and gladneſs of heart, being in better health, except ſometimes a common cold, or rheumatic complaints, than at any time of my life; I both ride and walk about with eaſe and pleaſure to myſelf; my complexion, from being pale and wan, is become florid; and, from being thin and emaciated, I am got plump: but I ſtill continue taking the medicine; for, as I have ſuffered ſo ſeverely, I cannot yet venture to leave it off. I began to take it in the ſummer of 1789, and I took a quarter of a pint-tumbler twice every day regularly till the 1ſt of January 1790; and as I was then much relieved, I took it only once a day till January 1791; and from that time I have omitted taking it one day in a week; and if I find I continue well, I ſhall omit it farther, from the begining of the next year."

CASE XXVII.

"In the year 1779 I was attacked with a ſtrangury and total ſuppreſſion of urine for ſeveral hours, without being able to aſſign any cauſe; but, after taking emulſions and mucilages, I was enabled to paſs urine again: I had many returns of the complaint; and in particular, one time I was adviſed to drink ſome gin and water, which I ſoon found to increaſe my pain, and cauſe greater irritation to make water. I was put into a warm bath, without any good effect, and a ſurgeon drew off my water by a catheter; and he ſo wounded the paſſage near the proſtate glands, that a great quantity of blood paſſed off with the urine. He told me I had a ſtone in my bladder: being rather alarmed, I then conſulted the late Mr. Elſe, who, on paſſing a bougie, told me my complaint was not calculous, but from a ſtricture. I alſo conſulted Mr. John Hunter, who likewiſe told me I had a ſtricture, and adviſed me to wear bougies: I did ſo; but not being able to indulge during my wearing them, they frequently irritated the diſeaſed part very much. Here I muſt obſerve, that, on my introducing the

the bougies, they have seemed to rub against some very hard rough substance near the neck of the bladder; but, on withdrawing them, I never could observe any impression had been made on them. During my wearing bougies, from 1779 to 1786, I had many total stoppages of urine, from which I could always relieve myself during the first part of the above period, by introducing a catheter; and afterwards the passage became too narrow for the smallest catheter I could get, and I made use of a small bougie. From 1786 I left off the use of bougies, and continued to make urine with difficulty; but had not any total stoppage until December 1790, when, having drank one evening rather freely of some very strong brandy and water, I had frequent irritations to make water, which I then passed with some difficulty; and the next evening being again engaged in company, I very imprudently retained my urine a long time (though irritated to pass it), until I felt the usual symptoms of suppression. I had no sleep during the night; and being obliged to go from home five miles, I went on horseback; and being a very wet and cold day, I felt a chill on my skin, made many attempts to pass my urine, but without effect. I returned home, ordered a warm bath to be got ready, and sent for a physician and a surgeon, who, by the use of the warm bath, bleeding, anodynes, by the mouth and clyster, and other remedies, relieved me from the most distressing painful spasmodic efforts to void my urine, during the suppression (which continued twenty-eight hours), I had ever experienced; when the urine began to flow by drops, and my bladder was emptied in about six hours. Here I must mention that different-sized catheters and bougies were attempted to be introduced, but without effect. At the end of a month from this time, I was recovered from the weakness and irritability consequent on the suppression of urine; when I found the stricture and irritability of my urethra in the same state as previous to this attack. By the suggestion of a medical friend, who was then at Bath, my urine was tried with paper, stained with litmus, and found to be surcharged with acid; and by his recommendation I began taking the Alkaline Mephitic Water (prepared according to the directions in Dr. Falconer's pamphlet); from which I soon found my urine pass with more ease, and the stream rather fuller. By repeated trials of my urine during my taking this water, I have found it has effectually prevented

my

my urine being acid (for the litmus teſt-paper is not altered in colour when dipped in it), which I, as well as the medical gentleman who attended me, think was the principal if not ſole cauſe of the pain and inconveniences which have at various times been the conſequence of the ſtricture.

"I continue to drink the water from $\frac{2}{3}$ to a pint in a day, and have the happineſs to think I now paſs my urine with as much eaſe, and nearly with as much freedom, as I ever did in my life. I abſtain from no food whatever; I drink mild beer and wine at dinner; and only avoid ſpirits, fruit, and acids. I have not ever found the Aqua Mephitica to diſagree with me. I have a good appetite, and am well in health.

"P. S. When I have been in London for a ſhort time, and have omitted to take the Water, I have felt a return of the ſtricture; but on going into the country, and again taking the uſual quantity of the Water, it has been always immediately relieved."

CASE XXVIII.

Extract of a Letter sent to Dr. Stonhouse, *by* Benjamin Colborne, *Esq. dated Nov.* 15, 1791.

" As you acquainted me you had received benefit by the use of the Mephitic Alkaline Water, and that you had kept a journal of your case, I should be obliged to you if you would please to send it me, as I presume you will have no objection to the printing of it for the public good. Dr. Falconer is now going to publish *another* edition of the Treatise on the Mephitic Alkaline Water: your case, therefore, with some others not inserted in the former editions, will be an acceptable addition.

" Soon after I had received the letter from Mr. Colborne, I drew him up my case, as follows:

1786.

" On October 5, having no suspicion of a stone, or any previous symptoms of it, except

cept a more frequent irritation to make water, which I attributed to my age (being then in my 71ft year), I voided a round ftone with little pain. This alarmed me; but, as I was in a tolerable good ftate of health, confidering fome infirmities, and no troublefome calculous fymptoms, I was unwilling to have recourfe to medicine.

" November 15—I voided three more fmall round ftones, one after another, at one time.

" December 11—Another about the fame fize.

1787.

" On January 1—I voided two fmall ftones.

30—A larger round one.

" February 3—A fmall round ftone in the morning—a large round one in the afternoon.

27—A large round one.

" March 9—A little ftone, not larger than a vetch.

10—Another, thrice as large.

23—A round ftone, the fize of a large pea, and very turbid urine after it.

"May 10—A small round stone.
 23—Two small round ones.
"September 6—A large round stone.
"Oct. 12—A large round stone.
 18—A middle-sized round stone.

"During this year I kept my body open by soluble medicines, such as manna with oil, small doses of rhubarb, and occasionally with castor oil; and when in pain, as I sometimes was, I took oil with liquid laudanum, or pills of solid opium, barley-water with gum arabic; and I drank Bristol water on the spot. —My diet, chiefly white meats, or fish; abstaining from every thing salt, and hard of digestion. My general breakfast and supper was half a pint either of milk or chocolate, and which I shall scarcely ever alter.

1788.

"January 12—A very large round stone, which passed with difficulty. From this time to April I was sometimes in pain, particularly about the neck of the bladder, and now and then made water with pain at the end of the urethra, and turbid urine. I took softening things occasionally, as barley-water
with

with gum arabic, and linseed tea sweetened with honey, &c.

" In the month of April I communicated these circumstances to Dr. Fothergill at Bath, being then on a visit to my son-in-law, Mr. Vigor, at Bathford, and in a good deal of pain, seemingly about the *right* ureter: he advised me to try the warm bath, which I did twice with a degree of ease: he advised me likewise to use anodynes freely, and strongly recommended the trial of the Mephitic Water. On April 13 I began to take it, and took at four times two quarts of it, which did not disagree with me: but as I soon returned to my living at Cheserel in Wiltshire, 22 miles from Bath, I could not procure any more till May 5, when I received an hamper with four bottles of it, of which I took regularly half a pint a morning, and the same at evening. On May the 21st I received from Bath two quarts more: the whole therefore I then took, amounted to no more than eight quarts. Finding myself *easier*, and a difficulty in getting a supply of the water fresh and fresh from Bath, which must be conveyed in bottles, I determined to desist from a medicine I could not procure without interruption, nor in perfection: I took *that* in quart bottles, but afterwards I had half-pint bottles made, each to hold only a single dose.

" June 10—I made urine tinged with blood, as I rode in my carriage, and a turbid urine, but nothing passed.

" From this time to Sept. 5 I was tolerably well, rarely in pain, and voided no stones: but this temporary relief I cannot attribute to the small quantity I took (and with interruptions) of the Mephitic Water at that time; the only apparent effect of which, as I then perceived, was, that it was rather too diuretic.

" September 5—Great pain in the night, for the first time, seemingly about the right ureter.

" Sept. 21 and 25—Returns of the pain, more or less violent, and generally in the night; sometimes sick with the pain, but not much so.

" November

" November 5—The night and next day in great pain. I fomented the parts with bladders of hot water, and supplied it well with oil and liquid laudanum, and took large doses of opium an hour or two before bed-time.

" Nov. 11, 24, and 29—Violent pain, seemingly in the right ureter; but as I had had no certain symptom of the stone for some time, nor passed any since January 12, I question whether this violent pain might not have been *spasmodic*, especially as I have been subject to dreadful spasms (or cramps) in my legs, and sometimes on the right side of my breast.

" December 2, 18, 19, 22, 28, and 31— Violent pain, seemingly about the right ureter, which would last for some hours, and go off gradually; leaving neither bloody urine, difficulty in making it, or perception of any stone passing: it might therefore be *merely spasmodic*, for the reasons I have suggested.

1789.

" January 4, 6, and 8—Great pain, and occasionally sick with it, but to such a degree as to vomit. I took folid opium, but not with the relief I expected.

" Jan. 10—Pain returned: I fomented the parts with bladders, filled with warm water; embrocated it with oil and laudanum, drank emulfions, emollient infufions of rad. alth. coltsfoot, &c.

" Jan. 23—Pain returned with great vehemence, and lafted longer than ufual. From this day I rarely have had any complaints of *that* kind; but now and then voided turbid urine; the fediment of which was fometimes fo hard at the bottom of the pot, as to require a fcraper to get it off.

" On Wednefday, June 3, I had a farther converfation with Dr. Fothergill, and told him that, as he had fome time ago recommended the Mephitic Water to me; and as, being then at my livings in Wiltfhire, I found it inconvenient to procure it, efpecially as it muft be brought in bottles, and not caring to be at the trouble of making it myfelf; I had by no means given it a fair trial. The Doctor was of opinion, that it would be right in me to renew it; efpecially as the

fummer

summer was advancing, and as I could have an uninterrupted supply from Mr. Becket, in Corn-street, Bristol, who makes it in the greatest degree of perfection. I then desired the favour of Dr. Fothergill, as he was acquainted with Mr. Colborne, to introduce me to him, that I might relate the whole of my case to him, and have his sentiments, whether he would have me enter on a regular course of the water. Accordingly I waited on Mr. Colborne, June 3, 1789: no one could be more obliging to me than he was; shewed me various experiments he had made on the human calculi; and gave me satisfactory reasons, approved by the Doctor, why I should *immediately* enter on such a course. He supposed I had some small stone, or calculous yellowish sand at the neck of the bladder, which had not come away, and which occasionally brought on pain and irritation to void frequent and small quantities of urine. He was so kind as to supply me with a few bottles of the water, during the three or four days I staid at Bath, and on *that* day and the next I drank half a pint in the morning, and the same in the evening, with two table spoonfuls of hot milk in each dose.

"From

"From June 3 to 25 I took it without any *vifible* effect, except being rather diuretic.

"On June 25, 26, 27, I made urine tinged deeply with blood on walking, but without pain.

"I then wrote to Mr. Colborne from Briftol Wells (where I now live), to know if he thought I might fafely perfevere in the ufe of the medicine, as I had for three days fucceffively made bloody urine, on no other motion than gently walking. On his anfwer in the affirmative, I continued it for fix months regularly.

"July 2—Urine again tinged with blood, and fome drops of pure blood (previous to it); but with little pain.

"From July 3d to the 15th, for feveral days there was fufpended in the middle of a glafs of urine, what appeared to *me* a kind of mucus.

"July 16—I voided a fmall fragment of a ftone,

a stone, very hard; as indeed were *all* the stones I had before voided.

" July 22—Another small fragment of a stone, which appeared like a small stone divided into two parts.

" From the 15th to the 22d, I had a little of the mucus suspended in the glass of urine. No stone passed from the 22d to the 31st of this month, nor any mucus appeared.

" August 1—Some mucus suspended. 6—A little yellowish sand, evidently of the calculous kind, appeared at the bottom of the pot yellowish: most of the remaining days of this month, either sand or mucus, or both, came away from me.

" From September 1 to no 16 sand, and, except for two days, no mucus.

" September 15 to 18, no sand, except one day only.

" September 19, voided half an hard small stone, like the half of a cockle-shell.

" From

"From that day to September 30, only a little mucus now and then; but on this day I was sick, and in great pain at the end of the *urethra*.

"From October 1 to 9, free from sand or mucus, and quite easy; but on that day I voided some harder sand than perhaps I ever voided before.

"From October 9 to 24, only a little mucus; and on that day a little sand, not so hard as in general.

"October 26—The splinter of a stone, which came away with some pain.

"To the 31st, quite easy.

"November—This whole month quite easy, no sand, and only now and then a very little mucus.

"December—Quite easy all this month.

"My complaints being seemingly removed, I *desisted* from taking the Mephitic Water.

1790.

"From January 1, 1790, to May 14, 1791 (a year and four months), I continued free from my complaints, except more frequent irritations to make water, than when in a state of health; nor had walking or riding any bad effect on me, even though I disused an *hollow* cushion, which I *before* used, to prevent the neck of the bladder from pressing on the seat.

1791.

"In the month of April, 1791, I was confined to my bed by a fever, and a very formidable cough, with large discharges of thick phlegm; during which time I had for several days an excruciating pain at the neck of my bladder, and such a pain in making water, as was very grievous indeed; scarcely many minutes without calling for the pot. During this terrible situation I drank three or four quarts of the common emulsion of the London Dispensatory in the four and twenty hours; soon after which I had two very large jagged stones, adhering to each other, came from me, with inexpressible torture, and bloody urine. Mr. Lowe, of Bristol, my surgeon,

surgeon, could scarcely believe they could have passed through the urethra.

" I then found I had acted very imprudently (and severely indeed I suffered for it) in leaving off the Mephitic Water for so long a time; during which interval these stones, I presume, had generated: whereas, in all human probability, had I omitted the Water for a few months, and then taken it once in a day only, I should have had no return of my complaints, no future generation of the stones.

" Five days after this, in the month of April, the soreness of the *urethra* was so great, that I voided my urine with an inconceivable degree of pain. As soon as the part was healed, after so great a laceration, I had immediate recourse to the Mephitic Water: a few days after taking this, I voided the fragment of a small stone with little pain, in the month of May; and on the 30th of May, another fragment of it: neither of them so hard as any of the former stones that had passed.

" On June 1st, I voided a rough stone, of a middling size, with some pain and blood.

" I continued

"I continued the Mephitic Water during the months of May, June, July, August, and September; was free from pain, or stone, but now and then (though seldom) a little yellowish sand, and a little suspended mucus.

"As I found it *in my constitution* rather more diuretic than formerly, and as I had been so long in a manner easy, I had determined *gradually* to omit it, and to return to it again after a short period.

"But, to my suprise, on October 12th last, I voided a small angular stone with very little pain; since which, to this present day, November 23d, I have been quite free from any complaint of the *calculous* kind.

"Query?—Was this *small angular* stone left behind in the bladder from June 1st, since which day none had passed till October 12; or was it generated under a course of the Mephitic Water, during the many months I took it?—It seems to me most probable, as it does to Dr. Fothergill, that the stone, being small, remained in the kidneys, or bladder, and was not generated during the course of the Mephitic Water."

Bristol Wells, *Nov.* 23d, 1791.

Copy of a Letter from John Ingen-Houfz, *Body Phyfician to their Imperial and Royal* MAJESTIES, *to* William Falconer, *M. D.*

"Dear Sir,

"You will always find me ready, both as a man and as a phyfician, to contribute, as far as lies in my power, to the relief of human mifery, and to fecond your difinterefted views, directed to fo laudable an end, as that of communicating to the public one of the moft valuable, and perhaps the moft beneficial remedy ever difcovered againft the moft excruciating of all difeafes, the Stone and Gravel: a remedy which, having been prefented to the world in the moft liberal way, as foon as it was difcovered, reflects immortal honour on that worthy and truly philanthropic man, Mr. Benjamin Colborne, the inventor of it.

"I will firft defcribe you fome cafes, which my learned friend Dr. Van Breda, phyfician at Delft, in the province of Holland, communicated to me in different letters, fince I made him acquainted with this important difcovery; after which, I will give you fome account of my my own cafe, and of fome others, which fell occafionally under my obfervation fince my prefent refidence in your happy Ifland.

" A youth about 15 years old, fubject from his childhood to fymptoms of the gravel, was all on a fudden feized, in the middle of June 1790, with a very acute pain in the left kidney, accompanied with an almoft total fuppreffion of urine: thofe fymptoms being by proper treatment much abated, a troublefome pain remained for fome days in the region of the left kidney, after which the pain defcended gradually lower and lower towards the bladder, where, in the courfe of a few days, it fixed itfelf, extending through the lower and left part of the abdomen. The pain being fixed at that place, was foon accompanied with a pain in making water, principally at the time when the bladder was nearly emptied. Dr. Van Breda, not doubting that fome gravel or ftone was formed in the left kidney, and that in its defcent through the left urethra it ftuck towards its orifice, which opens into the cavity of the bladder, gave him a pint of the Mephitic Alkaline Water daily, which contained one drachm of falt of tartar. In the fpace of four or five days the urine came forth in greater quantity, and with much lefs pain; his appetite, which was much impaired, as well as his general health, were both much mended.

" After having taken the medicine fifty days, he was reitored to perfect health, and left Delft; fince which time the Doctor has heard no more of him, and never was certain if he ever difcharged a ftone fince he began to take the Mephitic Alkaline Water.

" A man, aged forty-eight years, laboured under a difficulty of making water more than two years, which increafed to fuch a degree in July 1790, that he could pafs no urine but by drops; and at laft almoft none at all could be paffed, but by means of a hollow bougie, by which, befides fome urine, a very thick mucus was alfo difcharged. He fuffered, before the application of the bougie, the moft excruciating pain and tortures, in the continual ftraining to make water; and the introduction of the bougie brought but a temporary and imperfect relief. Dr. Van Breda thought that the ufe of the Aqua Mephitica Alkalina might do him fome good, and accordingly gave him a pint daily.

" The

"The patient complained, after taking the firft dofes, of fome pain in his belly, which however went off foon. The patient began in a few days to void more urine, but mixed with a prodigious quantity of tough flime, fomewhat refembling jelly, which funk to the bottom of the chamber-pot, and adhered fo ftrongly to it, that, after the urine was ftrained off, it did not fall out, although the veffel was kept inverted.

"The quantity of mucus which the patient paffed during the firft ten or twelve days was not lefs than a pint a day: the pain, which was very great during the time this mucus was paffing, decreafed gradually, in proportion as more urine was paffed along with it. After having taken the medicine twenty-five days, almoft every fymptom of the difeafe had left him, his urine became of a natural colour, and no flime was to be feen in it. He continued from this time the ufe of the Alkaline Water, but took only half a pint daily. After having thus taken thirty-two pints, he became perfectly free from every complaint, and his health continued good for five months; but in January 1791 he acquainted his phyfician, that fome flime began again to make its appearence in his urine, which being examined by the teſt of Mr. Colborne's blue paper, was found to have an acid predominant in it. Dr. Van Breda adviſed him, on this, to begin again a courſe of the Mephitic Alkaline Water, which removed in a few days every appearance of that kind. The patient ſtill continues to take one doſe every day, and has remained free from every morbid ſymptom to the preſent time, namely, March, 1791.

" Since the communication of the above caſe, Dr. Van Breda related to me the hiſtory of ſeveral other diſeaſes affecting the urinary organs, cured by the ſame remedy; among which I will deſcribe two of the moſt remarkable.

" A lady of a corpulent habit of body, and paſt the meridian of life, inclined to a dropſical diſpoſition. Her urine came in ſmall quantities, and was at laſt almoſt totally ſuppreſſed; ſcarcely any being ſecreted by the kidneys. In this alarming ſtate ſhe took the advice of Dr. Van Breda, who preſcribed for her the uſe of the Aqua Mephitica Alkalina. She had ſcarcely taken two pints of it, before ſhe found herſelf much relieved; the urine began to be ſecreted

secreted more and more copiously; and she was in a few days almost well; and her appetite increased. She continues still the use of the medicine.

" In a letter of the same physician, dated Delft, July 4, 1791, the following very remarkable case was communicated to me:

" A patient labouring under the stone applied to Dr. Van Breda, who advised him to the use of the Alkaline Water: by the use of which he soon began to pass a number of small stony concretions, generally about the size of a small cherry-stone, which were easily broken by the fingers, and proved to be laminated white covers, or shells, containing another kind of small stones, very smooth, brown, and much harder than their covers, and of different sizes, some being no bigger than a small pin's head; besides these laminated stones or shells, he passed also a great deal of the same kind of stones, already broken into two, three, or more pieces, and a proportionable number of the brown smooth stones; which it was evident had been, whilst in the bladder, shut up in the laminated white stones; of which covers or shells the laminated fragments were

evidently

evidently the broken remains. His urine became alſo charged with a whitiſh ſediment, which, not being diſſolved in urine, was only ſwimming in it as an heterogeneous matter, which was in appearance cretaceous, and of the ſame nature with the ſhells or covers of the ſmall ſmooth brown ſtones. At the time this letter was written, the patient had voided ſo many of theſe ſmall ſtones or nucleuſes and their ſhells, that, if they had been put together, would have been as large as a pigeon's egg. Dr. Van Breda ſent me ſome of theſe calculi. I was farther informed that this patient, after having paſſed a great quantity of theſe calculi, and of the apparently cretaceous matter, grew daily better, and was, October 13, 1791, quite free from pain in making water. Though his phyſician thought he was not yet radically cured, yet he himſelf was ſo far ſatisfied, that he left off taking the Water, contrary to Dr. Van Breda's advice. That phyſician found the white ſhells above mentioned diſſolved eaſily in the Aqua Mephitica Alkalina, but that the brown ſmooth kernels did by no means diſſolve ſo eaſily in the ſame Water; they however at laſt grew ſpongy in this Water.

" By examining the ſtones myſelf, I found that neither the white covers, nor the brown nucleuſes

nucleuses effervefcid, either with vitriolic acid, or with falt of tartar.

" On breaking fome of the hard brown kernels, and obferving them with a microfcope, I found they had about their centre a fmall fmooth cavity, in which, very probably, there had been another original nucleus, though I did not actually find it: or perhaps it flew away, or broke by the blow, by which I broke the brown kernel; or perhaps it efcaped my fight by its fmallnefs.

" The fame phyfician cured alfo lately a patient labouring under fevere rheumatic pains in his hands and feet, accompanied with occafional fwellings. Six bottles of the Aqua Mephitica Alkalina performed a complete cure. I will now give you a fhort account of my own cafe.

" After having paffed, fince the year 1780 (when I returned to Vienna, after an excurfion to France, Holland, and England), an almoft conftant contemplative and fedentary life, contrary to my former active manner of living, I found myfelf at laft afflicted (being then at Paris, 1788), almoft at the fame time, with both the gout and the gravel. The pain beginning in the left kidney, went down to the bladder, where a ſtone, half an inch long, and one ſixth part of an inch in diameter, remained for ſome days, and gave me very excruciating pain, principally in making water. I at laſt got rid all at once of theſe ominous ſymptoms, by paſſing the ſtone without any difficulty or ſtrain. Soon after this period I began to paſs now and then one or two ſmall ſtones, all very hard, reddiſh, and cryſtalliſed: my chamberpot was very often lined with a red ſandy fur, and I found myſelf at the ſame time afflicted with ſymptoms of biliary concretions. I got three or four times, in the ſpace of three or four months, the jaundice; which was always preceded by a very troubleſome pain at the upper part of the abdomen. I took the advice of ſeveral of my medical friends at Paris; but none gave me more ſatisfaction than that which I received from my old friend Count Carbury, a very learned man, and who for many years had been honoured with the place of body-phyſician to their Royal Highneſſes the Count and Counteſs d'Artois. He adviſed me to take every morning, two hours before riſing, the expreſſed juice of a whole lemon, ſweetened with ſugar or honey, and

mixed

mixed with about two chocolate-cups full of warm veal or chicken broth.

" The Count mentioned to me several patients who were cured by this remedy, among whom was the late Doge of Venice, to whom it was prescribed by the late celebrated Baron Van Swieten. I continued the use of this remedy for about eight months, and also took daily about four or five ounces of honey. Although I was not cured by the use of what I have just mentioned, yet as I had no return of the most painful and alarming symptoms, after having taken it some time, I have reason to believe that it had some salutary effect in checking the violence of such a complicated indisposition.

" I left off, indeed, with some regret, the use of the lemon-juice, when I began to take the Aqua Mephitica Alkalina; because I had some degree of confidence in the remedy, and partly because it was to my palate the most agreeable thing I ever tasted; exciting, besides its truly delicious taste, a most enchanting sensation when it reached the stomach, which, like a true *nepenthes Helenæ*, pervaded all my limbs, and produced a new and durable

rable sensation of the most pleasurable kind, and such as I could never have imagined was possible to take place. I am not, however, certain that the same sensations would be produced in every person by the same means. I continued, after this, the use of honey, but in less quantity. The first information I received concerning the Aqua Mephitica Alkalina was at Rotterdam, in the month of October, 1789, from my friend Dr. Becket, secretary to the philosophical society of that city; a truly learned man, as well as an excellent and successful practitioner.

" This gentleman lent me your book on the subject, and communicated to me several cases, in which the Aqua Mephitica Alkalina had been successful.

" I proceeded soon after on my journey to London; having never been, since I first perceived any symptoms of the stone, free, during a whole month, from passing some small gravel or sand, or from some uneasiness or other that denoted a calculous disposition. I was alarmed by new pains in the left kidney, when, travelling between Harwich and London, and the day after I arrived in that metropolis,

metropolis, I voided in the morning two small stones, very hard, and of a reddish hue, and composed of shining cryftals. The next day I found my chamber-pot lined with a red fur, sharp to the touch.

" I went the same day to visit my old and respectable friend Sir George Baker, physician to the King, who, on being made acquainted with my case, advised me to begin immediately a trial of the Aqua Mephitica Alkalina, and gave me on the spot a pint bottle of it, which he happened to have in his house. I took immediately half of it, and the remainder towards night, and ordered immediately some bottles to be sent to me from the shop to which Sir George Baker directed me; and have continued the use of it from that time to this day, taking regularly, every day, one drachm of salt of tartar, neutralized by Fixed Air, as in the Aqua Mephitica Alkalina. I take one half of my daily quantity about two hours before I rise, and the remainder on going to bed; and have the pleasure to inform you, Sir, that, since the very day I began the use of this remedy, I have remained quite free from every symptom of that dreadful disorder.

" My

"My health, which was not a little impaired by such a complicated indisposition, has been constantly improving; I have had no return of the gout, even in the smallest degree, and during a whole year no symptoms of biliary concretions.

"After this account, you will readily believe, that among those who owe their happiness and comfort to this remedy, and who ought to manifest a sincere gratitude towards the benevolent inventor, I must place myself in the first rank; and I should be unjust if I did not take this opportunity of acknowledging publicly what is due to you, Sir, as the principal and disinterested promoter of the use of this truly wonderful remedy.

"Before I finish this letter, I will give you some account of a few cases, to which I was myself a witness.

"A man, eighty years old, a common labourer, had been afflicted for the last twenty years with a calculous complaint in the bladder. Although he suffered a great deal, he had not been prevented from working so much as to procure him a livelihood, except during

during the two laſt years; in which ſpace of time he had remained almoſt in continual pain, eſpecially when making water; ariſing from a frequent and almoſt perpetual diſcharge of ſand, ſmall calculi, and ſharp urine. About the beginning of laſt ſummer he began to take the Aqua Mephitica Alkalina, and in a few weeks found ſo much benefit from it, that he could work again as before; his urine paſſed without trouble, and free from any calculous concretions. The large ſtone, which he has ſtill in his bladder, gives him pain, only at intervals, and in certain ſituations or poſtures of his body. This caſe, and ſome others, which fell under my examination, have ſuggeſted to me a more probable reaſon than the one uſually given, for the great relief which perſons labouring under the ſtone in the bladder have experienced from the uſe of the Cauſtic Alkali, or the Aqua Mephitica Alkalina, although the ſtone remained undiſſolved in the bladder.

" It has been ſaid by ſome, that the uſe of Alkaline ſubſtances (though experience has ſhewn ſuch effect to be contrary to their nature) produced upon the ſurface of the ſtone a covering of mucus; but I think it more rational

rational to fuppofe, that Alkaline fubftances produced this effect, by neutralifing the predominant acid acrimony of the urine, and preventing the farther concretion of calculus.

"A gentleman of my acquaintance, aged about forty, was troubled with difficulty and pain in making water, which was of fifteen years ftanding; the original caufe of which was an inflammation and abfcefs in the proftate gland from a venereal caufe: the difficulty in making water increafed at times, fo as to amount to a total fuppreffion, and greatly endangered his life. A hollow bougie, which however could not be introduced without great difficulty and pain, faved him more than once from death.

"The principal feat of his pain was about the neck of the bladder. An acid being difcovered to predominate in his urine, he was advifed to the ufe of the Aqua Mephitica Alkalina; this relieved him in a fhort time, fo much as to enable him to pafs his water freely, and with very little pain; though there is no doubt but that the proftate gland remains
ftill

still in a morbid state. He perseveres in the use of the remedy.

"I can assure you, that among those who have continued the use of the Aqua Mephitica Alkalina for a long time together, I have not observed that any indisposition whatever, which could be ascribed with any degree of probability to the use of the medicine, had taken place. On the contrary, the digestion and strength of the patients, which in many (among whom I may reckon myself) had been impaired by long and severe sufferings, have been in general remarkably improved.

"Several cases besides, in which a perfect cure was obtained, in complaints of the calculous kind, by the use of this remedy, have been communicated to me in different letters, since I dispersed the information concerning its efficacy upon the Continent, which I did in two papers inserted in the two first parts of a new chymical journal, published in Holland, and entitled, *Scheikundige Bibliotheck*. But as the particular circumstances of these cases were not accurately described, I can only say in general, that the use of this remedy prevails more and more abroad in proportion to its becoming more known, and its effects experienced. I leave you perfectly at liberty to make what ufe you think proper of this letter, and remain,

Yours, &c.

John Ingen-Housz."

Bath, November 25, 1791.

Many other accounts of the good fuccefs of the remedy have been received both by Mr. Colborne and by myfelf; but feveral of them were fo fimilar to thofe already related, that it was judged unneceffary to augment the bulk of this pamphlet (perhaps already too large) by the infertion of them, and others were not permitted to be authenticated with the names of the perfons who had received the benefit. It was at firft my intention to infert none but fuch as had the name of the perfon annexed, as a voucher for the truth of the narrative : this, however, is departed from in a few inftances; but in thofe I can teftify that the accounts came from authority that cannot be doubted, though it is not permitted to be vouched.

I would

I would farther mention, that I have been informed from the moſt unqueſtionable authority, that the Mephitic Alkaline Water has been of the greateſt ſervice in a caſe of violent ſtrangury, without any ſuſpicion of calculus, which returned about every ten or fourteen days. It prolonged the intervals to ſeveral months, abated the violence of the pain, and diminiſhed the heat of the urine.

Leſt it ſhould be alledged that the caſes above recited, however truly and candidly deſcribed, may, notwithſtanding, be a ſelection only from a number of others, in which this preparation may have been found unſerviceable or hurtful, I think it neceſſary to aſſure the public, that no caſe has fallen under my perſonal obſervation, wherein the Mephitic Alkaline Water has appeared to be in the ſmalleſt degree prejudicial; nor have I ever heard that it proved ſo from the report of others. One caſe only has occurred to me, wherein it was of no ſervice whatſoever; and in this the principal ſymptom was a frequent and painful urging to paſs the urine, which came away in ſmall quantities, but with little alteration in colour, ſave that a few ſpecks of blood were ſometimes viſible, but no gravel

gravel or mucus. As this feemed to be owing to the acrimony of the urine, the Mephitic Alkaline Water was advifed; but it was not fuccefsful, though it no ways aggravated the complaint.

EXPERIMENTS.

On the solvent Effects of the Alkaline Solution, saturated with Fixible Air.

By BENJAMIN COLBORNE, Esq.

A FRAGMENT of a calculus, of an ochrous colour, and rough on the outside like a mulberry, weighing fifty-one grains, was put into about two ounces and a half of the Mephitic Alkaline Solution, and corked up. After two days standing, the solution was poured off, and a fresh portion put on; and this was repeated every day, or every other day, for thirty-one days succeffively.

At the end of that time the stone was again weighed, and found to have lost thirty-six grains of its original weight.

Another fragment of the same calculus, weighing 41 grains, treated in the same manner, lost in thirty-seven days thirty-two grains.

Another fragment of the same, weighing fifty-four grains, treated as above-mentioned, lost in thirteen days thirty-two grains.

Another fragment of a calculus, of a light ochrous colour, and close texture, weighing forty-one grains, lost by the same treatment, in thirty-three days, eleven grains only.

A smooth white calculus was sawn into two pieces, one of which, weighing 29 grains, was put into the alkaline solution, but imperfectly saturated with fixible air; the other, weighing twenty grains, was put into an equal quantity of the solution perfectly saturated: after standing twenty-eight days, the first had lost six grains, the other eight grains.

A human calculus was divided into four parts; the first, No. I. weighing twenty grains, was put into the saturated alkaline solution, made of the common salt of tartar of the shops; the second, No. II. weighing nineteen grains, was put into a similar solution made with a proportionable quantity of the oleum tartari per deliquium; the third, No. III. weighing 18 grains, was put into an alkaline solution made with salt of tartar,

procured

procured from Apothecaries Hall; and the fourth, No. IV. weighing 18 grains, into an alkaline folution made with the cauftic lixivium, neutralized by means of fixible air, and as nearly as poffible of the fame ftrength with the others. After ftanding 45 days, No. I. had loft 13 grains; No. II. 13 grains; No. III. 14 grains; and No. IV. 11 grains.*

A piece of calculus, weighing 51 grains, put into the neutralized alkaline folution, made with lixiv. tartari, loft in 18 days 29 grains.

Another piece, weighing 56 grains, put into an alkaline folution made with foffil alkali in the fame proportion, and neutralized in like manner, loft in 18 days 13 grains.

Another piece, weighing 55 grains, put into a neutralized folution made with falt of tartar, loft in 18 days 11 grains.

Another

---

* *N. B.* Thefe different alkalies were tried, to difcover if one alkali had a greater power than another.

Another piece of calculus, weighing 41 grains, put into a neutralized alkaline folution, loft in 31 days 30 grains.

Another piece, weighing 49 grains, put into a neutralized folution made with falt of tartar, loft only four grains in the fame time.

A piece of calculus, weighing 56 grains, was put into a neutralized folution made with foffil alkali: in 31 days it loft 18 grains.

Another piece that weighed 64 grains, put into a folution of only half the ftrength, made with lixiv. tartari, loft in 31 days 42 grains.

The calculi above mentioned were corroded in holes like a worm-eaten piece of wood, but externally preferved their original figure, till they all at laft fell to pieces.

ADDITIONAL EXPERIMENTS.

By the Same.

Experiment I.

October 16, 1786. A fragment of a hard, close-grained human calculus, weighing fifty-five grains, was put into a large wide-mouthed vial, and upon it was poured *daily* the first urine that was passed, after taking a dose of the Mephitic Alkaline Water, by a person that was in a course of taking it every day. The vial was set in a moderately cool place, and the urine regularly changed.

	Loss of Weight.	Weight of the Remainder.
From the 16th of October to Nov. 16, it lost - - -	2 gr.	53 gr.
From Nov. 16, to Dec. 16	7 gr.	46 gr.
From Dec. 16 (1786), to Jan. 16 (1787) - - - -	10 gr.	36 gr.
From Jan. 16, to Feb. 16.	10 gr.	26 gr.
From Feb. 16, to March 16	4 gr.	22 gr.
From March 16, to April 16	4 gr.	18 gr.

Experiment II.

Another fragment of the same calculus was put into a wide-mouthed vial, and upon it was poured every day the urine of a healthy person who never had any signs of gravel, and who was not in the habit of taking any medicine whatever. The calculus weighed, when the urine was first put upon it, 45 grains.

	Loss of Weight.	Weight of the Remainder.
From October 16, 1786, to Nov. 16 - - - - -	0	45 gr.
From November 16, to December 16 - - - -	0	55 gr.

About the latter end of December, the urine was neglected to be changed, and the same urine remained upon the calculus until January 26th, in which time the fluid had become more fœtid and alkaline. The calculus had, during this time, fallen into three pieces, and had lost in weight ten grains. From that time the urine was changed regularly every day. On the twenty-sixth of January, the fragments of the above calculus weighed 35 grains.

	Lofs of Weight.	Weight of the Remainder.
From January 26, to Feb. 26 - - -	0	35 gr.
From February 26, to March 26 - - -	Gain of Weight. gr. 1 fs.	$36\frac{1}{2}$ gr.
From March 26, to April 26 - - -	$2\frac{1}{2}$ gr.	$37\frac{1}{2}$ gr.

Experiment III.

January 24th, 1787, an entire calculus, of a white colour, and fixty grains in weight, was put into a wide-mouthed vial, and on it was poured every day fome of the urine of a perfon who was in the habit of taking the Mephitic Alkaline Water in the fame manner as is mentioned in Experiment I. and the urine renewed daily. In the fpace of two months, the calculus was diminifhed in weight eight grains, and in another month the whole diminution was twenty-five grains. The laminæ that form the calculus alfo began to feparate; and it appeared, that the action of the folvent had penetrated much deeper in one part than another.

Expe-

Experiment IV.

A fragment of another very hard red calculus, which weighed fifty-four grains, was treated in the fame manner, and for the fame time as in the laft Experiment. It loft in that fpace of time feventeen grains.

Obfervations on the Symptoms attending Perfons afflicted with Calculus, and on the Effects of the Mephitic Alkaline Water.

By the Same.

Urine in general will change paper ftained with juice of turnfole to a red colour, which will be permanent; but the urine which is firft made after taking the Mephitic Alkaline Water, in thofe perfons who have taken it for fome time, will change the turnfole paper to a blue colour. This will take place even if the the Mephitic Alkaline Water be taken not more than a quarter of an hour before the difcharge of the urine.

If the Mephitic Alkaline Water be faturated with fixible air, it will not produce any immediate

mediate change on the turnfole paper; but after a fhort expofure to the air, the paper will become blue, as the fuperabundant quantity of fixible air flies off.

The urine of almoft every perfon in health, if fuffered to remain for twenty-four hours in the chamber-pot, forms more or lefs incruftation on the bottom and fides. This, however, I believe, never takes place in the urine of thofe who are in the habit of taking daily a competent dofe of the Mephitic Alkaline Water.

The urine of people fubject to the ftone or gravel is generally of an acid quality, and will then turn paper, ftained with the juice of turnfole, to a reddifh colour; and if fo, the perfon generally finds relief by the ufe of the Alkaline Water. But, if the urine turns the paper blue, it is moftly fœtid and putrefcent: in fuch cafes the Alkaline Water will take off the fœtor, and abate the general fymptoms, but muft not be depended on for a perfect cure.

Perfons fubject to the ftone or gravel, ought accurately to obferve the ftate of their chamber-pot, ber-pot, whether it keeps free of fur, or other adhefion to its bottom and fides; this being the principle criterion by which the increafe or amendment of the complaint can be afcertained.

If no difcolouration of the veffel appears after the urine has ftood in it for fome time, and particularly if the urine clears away any former adhefion, we may reft affured the urine is of a proper kind; but if the fides of the veffel grow foul, and this foulnefs accumulates, it indicates a ftate of the urine that tends to produce or increafe calculus.

Six or eight ounces by meafure of the Mephitic Alkaline Water, taken daily, will be found fufficient to keep the urine in a proper ftate by the generality of poeple; others may require double that quantity.

The effect of the Mephitic Alkaline Water in diffolving the incruftations formed by the urine, affords a ftrong prefumption in favour of its diffolving power on the calculus; therefore whoever voids any calculous fragments during the time of drinking the Mephitic Alkaline Water, has great reafon to think that they
are

are parts of an old concretion mouldering away, provided however a sufficient quantity of the remedy be taken to prevent any fur concreting on the chamber-pot.

As the Alkaline Mephitic Water is so efficacious in obviating the acrimony of the urine, it seems likely to be of service if given immediately after the operation of lithotomy has been performed; as it is well known that the healing of the wound is often much retarded by the irritation of that discharge, which is more likely to be troublesome in this way, as it is in its own nature more acrimonious.

Schirrosities of the os uteri and of the prostate glands, by retarding the passage of the urine, often produce symptoms similar to those that arise from calculus. One difference however may be remarked, which is, that people who have such schirri, bear the motion of a carriage or of a horse better than is done by those who have calculus; and if they void mucus, it generally comes away with the last drops of their urine, and the pain they feel lasts in much the same degree, during the whole of the time the urine is passing, which is seldom the case in calculous complaints; as the pain in them is generally moſt acute, juſt as the laſt drops are diſcharging.

People who have ulcers in the urethra attended with ſtricture, generally void purulent matter previous to the coming away of the urine, which laſt, by being long retained, ſometimes cauſes abſceſſes in the perinæum and ſuppreſſions of urine.

I have known perſons, of both ſexes, advanced in life, complain for many years of frequent urgings to make water, which comes away by little at a time, and is generally of a wheyiſh appearance, and, after ſtanding twenty-four hours, depoſits a large mucous ſediment. Sometimes a ſuppreſſion takes place. All the caſes of this kind in which the Mephitic Alkaline Water was tried, found more or leſs relief, one only excepted, which on examination proved to be a ſchirrus of the os uteri.

In violent paroxyſms of the ſtone or ſtrangury, I have adviſed the uſe of opiates combined with the Mephitic Alkaline Water. Fifteen or twenty drops of the thebaic tincture may be taken in a quarter of a pint of the water, and occaſionally repeated. A bag of

oats

oats heated in boiling water, I have experienced to be a convenient and fafe method of applying a fomentation to the os pubis, and what generally gives eafe. A clyfter alfo of two ounces of olive oil, and forty drops of the thebaic tincture, may be injected and retained for feveral hours. If neverthelefs the Mephitic Alkaline Water fhould prove two ftimulant to be repeated during the fit, which feldom happens, barley-water with gum arabic, may be fubftituted in its place; and when the pain has fubfided, recourfe may again be had to the Mephitic Alkaline Water.

The following Experiments were made by myfelf on the fame fubject.

Two calculi of a fimilar appearance, of a whitifh colour with a pink tinge, and of fuch a confiftence as to be eafily fcratched with the point of a knife, the one weighing five grains, and the other two grains and a half, were put into fix ounces of the alkaline folution, as above; in 38 days, during which time the folution was changed fix times, they were diminifhed in weight five grains and a half, but the apparent fize was little lefs than at firft;

they were however so friable as to fall to pieces on flight touching.

Two other small calculi, similar in appearance to the others, and both weighing six grains and a half, were treated in the same manner. In 38 days they were both of a scaly appearance on the outside, and of a hollow worm-eaten texture within, and withal so shivery, as to fall to pieces on slight pressure. The pink tinge on the outside was much diminished, but was retained within. The weight was only gr. 1 ss. so that they had lost 5 grains.

Six small calculi, similar to the foregoing, and weighing gr. iv. ss. were treated as above. In 38 days they had lost three grains and three-fourths, and were so fragile as to fall to powder on being touched.

COMPARATIVE TABLE

Of the solvent Power of the Alkaline Solution, saturated with Fixed Air, with Water simply impregnated with Fixed Air.

Mr. Colborne's Experiments with the Mephitic Alkaline Water.

Original weight of the calculi.	Time they continued immersed.	Weight lost by the calculi.
51 grains.	31 days	36 grains.
41	37	32
54	13	32
41	33	11
20	28	11
51	18	29
55	18	11
41	31	30
49	31	4
64	31	42

Mr. Colborne's Experiments with the Urine of a Person who was taking the Mephitic Alkaline Water.

Original weight of the calculi.	Time they continued immersed.	Weight lost by the calculi.
55 grains.	182 days.	37 grains.
60	90	25
54	90	17

EXPERIMENTS I made myself with the Mephitic Alkaline Water.

Original weight of the calculi.	Time they continued immersed.	Weight lost by the calculi.
$7\frac{1}{2}$ grains.	42 days.	$5\frac{1}{2}$ grains.
$6\frac{1}{2}$	48	5
$4\frac{1}{2}$	48	$3\frac{1}{2}$

EXPERIMENTS made by Dr. Percival on the dissolvent power of Water, simply impregnated with Fixible Air, on human Calculi.

See Percival's works, Vol. III.

Original weight of the calculi.	Time they continued immersed.	Weight lost by the calculi.
152 grains.	2 days.	$2\frac{1}{2}$ grains.
$165\frac{1}{2}$	2	11
$126\frac{1}{2}$	2	gr. $\frac{1}{2}$
$68\frac{1}{2}$	2	$3\frac{1}{2}$

Experiments I myself made on the solvent power of Water, simply impregnated with Fixible Air, upon human Calculi.

See Experiments and Observations on Fixible Air, London, printed 1776.

Original weight of the calculi.	Time they continued immersed.	Weight lost by the calculi.
6 grains.	15 days.	$4\frac{1}{2}$ grains.
7	31	4
$4\frac{1}{8}$	31	$2\frac{1}{2}$
5	31	$2\frac{1}{2}$

EXPERIMENTS

On the antiseptic Qualities of the Alkaline Solution saturated with Fixible Air.

Three pieces of lean mutton, a drachm each in weight, were, on Dec. 21, 1784, severally put into eight ounces of spring water, into the same quantity of water saturated with Fixible Air, and into the same quantity of the Alkaline Solution, and all closely corked up, and placed in a room wherein a constant fire was kept. The weather being very cold, no change was perceived for several days.

On Dec. 29, the vial with the simple water began to look cloudy, but scarcely any smell was perceivable.

The others continued clear and sweet.

On Jan. 2d, 1785, the smell was more perceivable, but still faint, in the vial with simple water; some little of a musty smell was perceivable in the vial with water saturated with Fixible Air; but the Mephitic Alkaline Solution still continued free of smell, and the fluid clear.

Jan. 3d. The vial with the simple water had acquired a smell evidently putrid. That with the water simply with Fixible Air had the musty smell much increased. That with the Mephitic Alkaline Solution was perfectly sweet.

EXPERIMENTS

To determine the comparative Quantity of Fixible Air contained in Salt of Tartar, and in a proportionable Quantity of the Mephitic Alkaline Water.

EXPERIMENT I.

Two ounces, by meafure, of the Mephitic Alkaline Water were put into a vial about three ounces contents, and exactly counterpoifed in a nice balance. This with the correfponding weight being fet afide, I mixed forty drops of oil of vitriol with one ounce, by meafure, of water, and counterpoifed that alfo, together with the vial that contained it. I then added the acid fluid to the Alkaline, drop by drop, until all effervefcence ceafed; after which, I again weighed both the vials with their refpective contents.

The acid liquor had loft one hundred and fixty-feven grains and the Alkaline had gained only one hundred and fifty-eight; fo that nine grains of air were diffipated. Hence it appears, that fifteen grains of Salt of Tartar in folution, was capable of containing nine grains of Fixible Air, which, if we fuppofe Fixible Air to be be in weight to common air as three are to two, amount to about thirteen ounce meafures.

Experiment II.

Twenty grains of dry Salt of Tartar were put into a vial, and accurately counterpoifed, as was another vial containing diluted vitriolic acid; I dropped the acid upon the alkali until all effervefcence ceafed. On weighing each of them again, the acid was found to have loft fifty-three grains, and the alkali to have gained forty-feven and a half—difference, five grains and a half. The dry Salt of Tartar therefore appears not to be half faturated with Fixible Air, being in this refpect to that contained in the Mephitic Alkaline Water, as eleven to twenty-four.

According to Dr. Dobfon's Experiments on this * fubject, the Salt of Tartar he ufed contained a fmaller proportion of Fixible Air than that here employed, two drachms being computed by him to contain only twenty-eight grains of Fixible Air, whereas, in the above experiments, that quantity is reckoned to contain thirty-three grains.

* See his Medical Commentary.

REMARKS UPON,

AND

INFERENCES FROM,

THE FOREGOING

CASES AND EXPERIMENTS.

THE Cases above related, which are all described, either by the parties themselves, or from the accounts of professional persons who attended them, whose veracity is unquestionable, will no doubt have their due weight with the reader.

Case I. exhibits an instance of a complete cure performed on a person considerably past the meridian of life, who had been for eighteen years afflicted in the most grievous manner with this complaint, and who had tried almost all the boasted remedies, without their affording any abatement of his sufferings, and with manifest injury to his general health. In this instance, the Mephitic Alkaline Water seems not only to have proved a specific remedy

medy for the calculous complaint; but also to have repaired, to a degree far superior to what could have been expected, the injuries done to the constitution both by the fatigue and distress incident to the disorder, and by the violent remedies which were used previous to his becoming acquainted with the efficacy of the Mephitic Alkaline Water. Time, the only test of truth, in such instances, seems to have ascertained the fact of his recovery beyond any possibility of doubt. Upwards of thirteen years have now elapsed since he began to make trial of this remedy, and during that time he has had no other interruption of ease than what might, as clearly as we can trace any occurrence in medical practice, be ascribed to the omission of the remedy before the tendency of the system to generate calculus was subdued. That time, however, appears to be now arrived, as far as such a thing can be determined; he having been able to lay aside the use of the remedy for several months together, without enduring any symptoms of his former complaint, which had before recurred with considerable violence on a much shorter interruption. I am happy, that at the interval of three years I am able to repeat the account given in the last

last edition of this work of Mr. Colborne's state of health. That gentleman's appetite, strength, complexion, and spirits, bear the fullest testimony to the innocence, if not to the good effects of the Aqua Mephitica Alkalina on the health in general; and the returns of the gout, to which he is constitutionally subject, have not been more frequent or violent than what might appear to be natural; nor has this disease attacked any of the vital organs, or caused any contraction or swelling of his limbs, the perfect use of which he now fully enjoys.

Case II. not only confirms the accounts of the efficacy of the remedy, in relieving the painful symptoms, but affords the greatest presumption that it possesses qualities of a highly solvent nature.

Mrs. Southcote had discharged numerous calculi, and several of a size to give exquisite pain in their passage, and had the greatest reason to think, from her own sensations, that one remained behind too large to pass; yet on a careful examination of her body, after her decease, which last was occasioned by a disease altogether unconnected with calculus,

culus, no calculus was found either in the kidneys or bladder, both of which were found to be in a perfectly found state; a circumstance scarcely to have been expected, even supposing no calculus to have been present, if we reflect on the dreadful sufferings which she had formerly undergone from the diseased state of those organs.

Case III. requires very little comment, as it exhibits a full and even minute account of a complete recovery, from as painful a state as we can well conceive to exist. It is worthy remark here, that the tendency of the system to generate calculus seems nearly, if not altogether, subdued, as appears from his being now able to intermit with impunity the use of the Alkaline Water for a long time together.

Case IV. is a notable instance both of the efficacy and innocence of the remedy, and indicates, as well as the foregoing cases, that the tendency to generate calculus may, by the long continued use of the Alkalina Water, be subdued.

Case V. is a remarkable example of the powers

powers of the Mephitic Alkaline Water, in a habit of body strongly disposed both to generate and to accumulate calculous concretion.

Case VI. evinces in the most satisfactory manner the safety, as well as the efficacy, of the remedy. A person of 84 years of age commenced its use, and continued it for three years without the smallest inconvenience to his health, and with the general alleviation of all his complaints.

Case VII. is an instance of the speedy relief which is afforded by the use of this remedy, and of its actually possessing powers of dissolving the calculus, or at least of diminishing the cohesion of its parts.

Case VIII. is a remarkable instance of relief being gained, when the organs that secrete and transmit the urine were probably in a very diseased state. Though no cure was alledged to be performed, and indeed not, in all probability, was any cure possible to be effected by any means; the symptoms that had been the most distressing, particularly the fœtor of the urine, were so much abated as to give

give little uneasiness in comparison with what had been before endured. Had the remedy been discovered earlier, it is probable, from the effects mentioned to be produced by it in Cases XVI. and XXVII. that it might have wrought a perfect cure.

Case IX. is an instance where a great temporary advantage was gained; but his complaint proving at last to be an ulcer of the bladder, no unlikely thing to take place, after a repetition of the operation of lithotomy, the Alkaline Water, though it afforded much alleviation, was unable to work a cure.

Case X. proves that in some cases a very weak solution of the Alkaline Salt, and taken but in small doses, may prove an efficacious remedy, even in a case wherein all the symptoms were of the most painful and urgent kind.

Cases XI. XII. XIII. XIV. afford the most satisfactory proofs of benefit received in such disorders; but are nowise particular, save that in one of them (Case XII) all the uneasy symptoms were removed, although there

there was every reaſon to think that a calculus remained.

Caſe XV. affords a remarkable proof of the efficacy of the Mephitic Alkaline Water, in a complaint of the urinary paſſages reſembling calculus, which was originally owing to external violence.

Caſe XVI. The perſon whoſe caſe is here deſcribed, appears to have laboured under a highly diſeaſed ſtate of the urinary paſſages, and perhaps of the ſecreting organs; yet theſe complaints have been totally removed, and the general ſtate of health, as we have every reaſon to think, much improved.

Caſe XVII. is in a good meaſure ſimilar to that immediately preceding.

Caſe XVIII. This caſe varies from moſt of the others, in that the relief gained thereby was ſlower acquired than in the others. The cure neverthelefs appears to have been as complete and as permanent as could poſſibly be expected; a circumſtance ſomewhat (at firſt ſight) extraordinary, at ſo advanced a period a period of life, were it not inftanced in other cafes. An important practical caution refults from the confideration of this cafe, which is, that thofe fo afflicted fhould not defpair of relief, and even of cure, even though the painful fymptoms fhould not abate fo foon with them as they appear to have done in moft of the hiftories here related.

Had the gentleman who is the fubject of the narrative here under confideration, not been poffeffed of great patience and fteadinefs, he would have failed of a cure, and his cafe might have been adduced as an inftance of an *unfuccefsful* trial of the remedy.

Cafe XIX. is a fair inftance both of the efficacy and innocence of the remedy; but affords occafion for no particular remark, fave what had been before obferved, that an advanced age is no objection, even in the fmalleft degree, to the trial of the Alkaline Water.

Cafe XX. befides exhibiting an inftance of the greateft relief being afforded in a truly calculous cafe, fhews alfo that the Alkaline Water

Water possesses a power of dissolving calculus.

Case XXI. affords a satisfactory instance of the efficacy of this remedy, after most of the things usually administered (soap lees particularly) had been tried without affording even temporary relief. The benefit this gentleman has received does not appear to have been influenced or obstructed by change of climate, sea voyages, or the attention requisite to an active military employment.

Case XXII. contains nothing particularly worthy notice, save the shortness of the time in which relief was procured.

Case XXIII. This case, as well as the foregoing, was communicated by Perry, an eminent surgeon of this city. It shews in the most satisfactory manner that the Alkaline Water possesses a power of dissolving urinary calculus. What degree of a lithontriptic it is endued with, does not appear; but we are informed that the solution of the calculus took place in less than twelve months.

It seems highly probable from the last account,

count, which ſtates that the patient has continued well, notwithſtanding his leaving off the Alkaline Water for three years, although the whole time he uſed it did not exceed one year, that the diſpoſition to generate calculus is ſooner overcome by this remedy in young ſubjects than in thoſe who are farther advanced in life.

Caſe XXIV. ſhews the efficacy of the Aqua Mephitica Alkalina, in a complaint evidently ariſing from the nature, and probably from the ſtimulant qualities of the urine.

The repeated good effects that followed the taking the medicine, and the return of the complaint on its being laid aſide, amount to as full proof that ſuch relief was owing to the remedy, as the ſubject is capable of affording.

Caſe XXV. which the reader can ſcarcely fail of obſerving to be drawn up with great accuracy and propriety, is a notable inſtance of the efficacy of the remedy, which does not appear to have been weakened in its efficacy, or to have had the benefit ariſing from its uſe,

use, protracted by the exercise of travelling, both on horseback and in a carriage, which was necessary in an active business.

Case XXVI. is perfectly satisfactory with respect to the benefit received in a very severe complaint of the calculous kind, which appears to have been hereditary in the family.

Case XXVII. is of a disorder of the urinary passages, attended with great pain and stricture, probably some disease of the prostate gland, but probably without any formed calculus. This case seems to evidence clearly, that the pain was caused by the acrimony of the urine, and the disease kept up, and probably produced at first by it. When this acrimony, which evidently appears to have been of the acid kind, was neutralized by the Alkaline Water, both the pain and stricture abated, and returned when that was omitted.

This case is important, as it proves that the Alkaline Water is capable of giving permanent relief, if not of effecting a cure, which is very probable, in complaints of the urinary passages, not proceeding from calculus, provided

vided a trial be made before any irreparable injury be done to the parts.

Cafe XXVIII. is a very circumftantial narrative of the cafe of a gentleman well known in the medical world, and who is now far advanced in life. The cafe is evidently calculous; and the efficacy of the Mephitic Alkaline Water is proved beyond a doubt, by the abfence of the painful fymptoms during its ufe, and their recurrence when it was laid afide.

The laft Cafe here adduced, is that of Dr. Ingen-houfz, contained in a letter to me from that gentleman, which contains many interefting remarks on its nature and efficacy, not only in calculous cafes, but in other diforders in which no trial has been made of it in this country that I know of. I am peculiarly gratified in being able to confirm from my perfonal acquaintance the account this learned and worthy phyfician gives of his own ftate of health, which feems as happy as an advanced life feems capable of enjoying. It muft afford comfortable reflections to perfons afflicted with calculous diforders, to find from the above Cafes, that, although the

Aqua

Aqua Mephitica Alkalina undoubtedly has a power of diffolving the calculus, and thereby ftriking at the root of the complaint; yet, that relief may be expected before any material diffolution of the calculus can take place, and even whilft we know it is actually fubfifting. Several of the Cafes fhew this, which can only be afcribed to the change produced by the remedy in the nature and qualities of the urine itfelf.

It is obferved of this difcharge, that in calculous paroxyfms, efpecially if accompanied with great pain, it is almoft always cauftic and irritating, like other fluids fecreted from inflamed parts. The mucus of the nofe, which is in general mild and bland, becomes frequently, by a catarrhous inflammation of the veffels that fecrete it, fo acrid, as to excoriate thofe parts of the nofe and lips upon which it falls. A fimilar change takes place in the urine, which, under fuch circumftances, generally feels fcalding and painful to the ducts through which it paffes; and this irritation conftitutes no fmall part of the mifery of the fufferers.

I would not, however, by any means deny that the mechanical action of calculous fubſtances is often ſufficient to cauſe great pain. Experience proves that this is frequently the caſe; but it is equally certain, that large calculi both of the kidneys and bladder have remained there many years with little trouble or uneaſineſs, and that even the pain produced by paſſing them is by no means proportioned to their ſize. A pretty large concretion, compared with the diameter of the urinary ducts, is mentioned, in one of the Caſes above recited, to be diſcharged without pain; whilſt others of a leſs bulk were often accompanied, in their paſſage, with great torture, and large effuſions of blood.

The particles of ſand, that come away, are often too inconſiderable to cauſe the uneaſineſs that is experienced, were not the membranes that line the ducts in a ſtate of inflammation, and conſtant irritability. This acrimonious condition of the urine is almoſt conſtantly accompanied with a diſpoſition to precipitation. Hence the turbid appearance of this diſcharge in ſuch paroxyſms, which the ſufferers often vainly flatter themſelves to be the criſis of their diſorder, when in reality it is no more than an indication of its prevalence. Both theſe circumſtances the ſaturated alkaline ſolution

lution is very efficacious in removing, neutralizing as it were the acrimony of the urine, and reftoring to it, together with its natural colour, its power of retaining in perfect folution thofe fubftances which it was intended by nature to difcharge.

Another circumftance much in favour of a trial of this remedy, is, that it acts without any violence of operation. The firft effects, obferved in all the inftances above related, feem to have been the abatement of the pain and uneafinefs, and the reftoration of the urine to its natural colour and other properties. It is found to act but mildly as a ftimulus on the urinary fecretion; and though in one cafe it may appear to have exerted fome aperient effect, this was fo inconfiderable as to render it a matter of doubt if it was to be imputed to the qualities of the medicine, or to the taking in an additional quantity of watery fluid, which, it is well known, will often produce that effect. It no where appears to have injured the appetite, digeftion, or general health. It has manifefted no feptic qualities in itfelf, nor produced any upon the fyftem; nay, thofe which took place from the ufe of the cauftic alkaline lixivium, ceafed during the trial of this

this remedy. The persons I have seen, who tried it, have exemplified its innocence respecting the general health, as strongly as its particular efficacy in this complaint.

It appears that the use of this medicine is not necessary to be superseded by slight indispositions. It has been taken in place of the common saline draught, and no very observable difference found in the effect; and one of the Cases shews, that it may be continued, without any apparent injury, during the course of a common gouty paroxsym.

I would not, however, assert, that the indiscriminate use of this remedy is admissible in all states of health. The quantity may often be an objection to some; the taste may prove disagreeable, and perhaps in some complaints (though I know of none at present) it may be specifically injurious. Experience, joined with prudence, is the only guide we have to direct us in such circumstances.

But although I think it probable, that the principal advantage derived from this remedy is owing to the change it produces in the urine; yet the experiments shew, that it possesses considerable powers as a solvent of the calculus. That its efficacy in this point of view may be compared with that of water simply impregnated with Fixible Air, I have formed the foregoing comparative table of their effects respectively. The difference in their solvent powers is inconsiderable; whilst the operation of the alkaline saturated solution is much milder, and, as I think, of a different kind from that of the simple impregnation of water with Fixed Air.

From examination of the effects of the two menstrua, it should seem that the action of the latter was principally upon the animal gluten or mucus that connected the sandy particles, which it gradually disunited, until they fell into powder; whereas, from the corroded and worm-eaten appearance of the calculi immersed in the alkaline solution, the sandy particles themselves seem to have been acted upon. How this is brought about, is matter of difficult investigation.

From Bergman's experiments, it appears, that the acid of sugar and calcareous earth, which probably form the stony part of the calculus, bear a stronger attraction to one another

another than any body does feparately to either; fo that the addition of no fimple fubftance, at leaft any that we can introduce into the body, will feparate them. But we fhould confider, that many bodies are capable of decompofition by a double elective attraction, that are not fo by any other means. Thus vitriolate tartar may be decompofed by folution of filver, though neither of the feparate ingredients would have any effect. This may poffibly take place here, the alkaline falt attracting the acid of the fugar, and the Fixible Air the calcareous earth; and as the former of thefe compounds is foluble in a watery fluid, and the latter fo when the Fixible Air is redundant, this may account for the clearnefs of the urine, and its freedom from precipitation, which the taking this remedy induces. The compound of the acid of fugar with calcareous earth is fcarcely foluble in water.

I make no doubt that the change in the *qualities* of the urine may be in part owing to the fame caufe. It is true, the faline fubftance formed by the union of the acid of fugar with calcareous earth, does not appear very acrimonious to the fenfes; but we fhould confider, that our fenfes are very imperfect judges of fpecific ftimuli.

ſtimuli. Tartariſed antimony and calomel, whoſe operation on the ſtomach and bowels is ſo violent, betray no ſuch effects in their ſenſible qualities; and we frequently find that clear, pale, and inſipid urine is retained with greater difficulty than what appears much more ſaline and acrimonious. It is poſſible that this compound may poſſeſs ſome ſpecific ſtimulus on the bladder and urinary organs.

The alkaline ſolution has exerted various degrees of a ſolvent power upon the different calculi; ſome reſiſting its operation more than others; but none have totally withſtood its influence. This difference may be owing to ſeveral cauſes; ſuch as the calculus having remained expoſed a longer time to the air, which increaſes its hardneſs, as it does that of ſeveral kinds of ſtone; its having been ſlower or quicker produced; or its containing a different proportion of animal mucus; and probably other circumſtances, which we do not at preſent, and perhaps never may, underſtand.

It appears pretty plain, I think, that diuretic remedies, merely as ſuch, have no good effects in calculous complaints. Independent of their ſtimulus, which I believe always to be be injurious, it is found, that a quick secretion of urine has no effect in preventing the generation of calculi. A gentleman whose case is related above, had a stone generated evidently during a course of the Harrowgate waters, which acted powerfully as a diuretic.

The experiments made with the saturated alkaline solution, as an antiseptic, confirm the observations of Mr. Colborne, on the effects it shewed on his urine; and we may infer from both, that no danger is to be apprehended from any putrefactive tendency, which, as an alkali, it might be supposed to produce.

In the former editions of this work it is mentioned as a probable conjecture, that a solution of the *fossil* alkali saturated with Fixible Air, might prove equally efficacious with the *vegetable* in the relief of calculous complaints. Mr. Colborne's very judicious experiments, which shew that it possesses the same solvent powers upon the gravelly concretions out of the body, together with the great similarity it bears in its chemical properties to the vegetable alkali, first led me to adopt this opinion, which has been since confirmed, by considering

what

what did not before occur to me, that this remedy in form of a mineral water had long been in use, and even high reputation, for the cure of similar complaints.

The waters of Carlsbad in Bohemia, so called in honour of the emperor Charles IV. who in the year 1370 discovered their medicinal virtues, are celebrated by Hoffman for their good effects in calculous complaints. " In cases,"* he says, " where the kidneys, ureters, and bladder, are clogged with tartarine mucus, and gravel, or where a calculus is actually formed, and by remaining fixed in the urinary passages excites

---

* Sive enim renes, ureteres et vesica muco tartareo et fabulo obsideantur, sive calculus et lapidosa materia in ureterum cavo subsistat et diros dolores suscitet, tam praeclara Carolinarum est facultas, ut copiosius ad renales tubos delatae, intus contentas sordes et materias alienas aliquando et fluxiles reddendo ejiciant, spasticis autem stricturis ureterum resistendo, eosdemque ampliando et laxando humoris que inhaerescentem in eorum alveo lapidem protrudant et elidant. Quin ipsam autem generationem calculi ejusque incrementum antevertunt; dum humores diluunt et salsedinem et acrimoniam eorum contemperant, quo minus mucosae et salsae materiae coire et concrementum lapidosum exoriri possit. Accedit et illud quod thermales hae aquae si urina cum ardore et dolore stillet, exoptatissimam spondeant opem. *Hoffman de Thermis Carolinis,* § XIV. Cap. V.

cites the moſt direful agonies, the virtues of the Carlſbad waters are particularly ſerviceable, by loofening the adherence of ſuch matters to the urinary ducts, and waſhing them away, and alſo in abating the ſpaſmodic ſtrictures of the paſſages, and thus procuring a free diſcharge for the calculous concretions."

" They moreover," as he ſays, " prevent the generation or increaſe of calculi, by their diluting qualities, and by their moderating the ſaline acrimony of the humours, by which the ſaline and mucous matter is prevented from acquiring a hard or ſtony conſiſtence. Theſe waters likewiſe produce the happieſt effects in abating the heat and pain that accompany the paſſage of the urine."

The good effects above mentioned to be produced by the Ca rlſbad waters, are almoſt exactly the ſame with thoſe which proceed from the uſe of the Mephitic Alkaline Water, which is by no means extraordinary, if we conſider that the two remedies reſemble one another ſo nearly. The water of Carlſbad contains, as its

principal

principal impregnation, the * foſſil alkali largely combined with † Fixible Air, which explains its beneficial operation in this diſorder more ſatisfactorily, than by referring it altogether to the aerial impregnation, as is done by Dr. Dobſon. However, according to Hoffman's account, it contains but a dilute ſolution of the alkaline ſalt, not more than one drachm ‡ being contained in two quarts of the water, whereas eight times that quantity is contained in the mephitic alkaline liquor. But we know that a conſiderable proportion of ſaline

---

* Affuſo ſyrupo violarum thermæ hæ viridem colorem induunt. *De Thermis Carolinis*, § VI. Cap. II.

† Thermæ hæ cum quocunque acido, ſive ſit mite, ut acetum, ſive forte, ut ſpiritus ſalis, vitrioli, aut nitri, effervefcunt cum magnâ bullularum et exhalationum in aerem copiâ. *De Thermis Carolinis*, § VI. Cap. II.

Licet odor circa ſcaturiginem obvius, volatile quoddam principium ſalinum prodat; non tamen urinoſi quippam redolet, ſed ſimilis odor ferme eſt illi, quando coquitur ſal tartari cum ejus cremore miſtum in aquâ. *De Thermis Carolinis*, § XIII. Cap. II.

‡ Ex libris duabus medici ponderis obtinuimus drachmam materiæ ſalino-terreæ, quæ aquæ ope iterum liquata, et per chartam emporeticam trajecta, concretione facta, ſalis puri dedit drachmam circiter dimidiam: hoc ſal, teſte ſapore, proxime accedit ad ingenium ſalis tartari intenſè alcalizati. Confligit enim cum quovis acido, et cum ſale ammoniaco mixtum, penetrantiſſimum volatilem ſpiritum elicit. *De Thermis Carolinis*, § VII. Cap. II.

line matter is always loft in fuch experiments, it not being poffible to recover from a folution of this kind, as much of the falt as we are affured it contains; part of it being diffipated by being changed into volatile alkali, by being united with phlogifton, and part loft by cryftallizing on the filtre, and by other inaccuracies unavoidable in fuch trials. It is therefore certain that the Carlfbad water contains a larger proportion of alkali than is ftated by Hoffman, though at the fame time the impregnation is not ftrong. To make amends however for its weaknefs, the Carlfbad water is drunk in much larger quantity than what the mephitic alkaline water is taken. Hoffman fpeaks of from 15 to 18 cups, as the daily quantity for rather* weakly people to begin with, which, fays he, is increafed by moft of thofe who ufe it to thirty, and by fome few to forty cups a day.

How much the cup or *ollula*, as he calls it, might contain, according to our meafure, I cannot determine; but I think we

can

---

* Proinde tutius eft primo die XV. tantum vel XVIII. ebibere ollulas, nifi majorem dofin firmitas, minufque mobilis corporis conftitutio fuadeat. Infequenti vero tempore, plurimis ad XXX. ufque quotidie ollulas, paucioribus ad XL. afcendere conducit. *De Thermis Carolinis*, Cap. VII. § VI.

can scarcely suppose that a mineral water could be given out in cups of less than a quarter of a pint contents. The middle dose then, at this rate, must be three quarts and three half-pints, daily, which would contain, according to Hoffman's calculation, which is unquestionably below the mark, 112 grains of the alkaline salt, which is not very different from the quantity contained in a pint of the Mephitic Alkaline Water above described, which holds 120 grains, which is to the quantity daily taken in the Carlsbad waters as 15 to 14.

The same celebrated writer ascribes nearly the same virtues to the Selters water, " which," according to his account, " has a wonderful efficacy * in complaints of the kidneys, bladder, or ureters, when these organs are either obstructed by tartarine mucus, or calculous matter, or in a corroded and ulcerated state. Whilst it abates the acrimony of the humours, it

---

* In renum, vesicæ et ureterum morbis, qui vel a tartareo muco et calculosâ materiâ has vias obstruente, vel ab erosâ et exulceratâ ipsarum substantiâ proveniunt, admirabilem habet virtutem. Dum enim humorum acrimoniam temperat, mucum dissolvit et eluit, renesq; vesicam et ureteres a materiâ ipsius inhærente repurgat, non modo ad calculum præcavendum, aut jam

'it dissolves and washes out the mucus, and clears the kidneys, ureters, and bladder, from any matter of this kind that may be lodged in them, and tends, not only to prevent the generation of a calculus, or to stop the increase of one already formed, but also affords the most desirable relief in the strangury, and when the urine is voided with pain and difficulty, and is thick and turbid in its appearance." The Selters water, as well as that of Carlsbad, is impregnated with the *fossil alkali, but in larger proportion, two quarts of the Selters water containing four scruples of alkaline salt, whereas the same quantity of the Carlsbad water contains no more than three. The alkali however here, as well as in the other, appears to be fully saturated

---

jam præsentis incrementa impedienda, præsentis est efficaciæ, verum etiam in stranguriâ difficilique ac dolorificâ urinæ crassæ ac turbidæ mictione, exoptatissimas affert suppetias. *Hoffman de Elementis et Viribis Fontis Selterani,* § XIV. Cap. II.

* Selteranarum libras duas medicas super leniori igne, quem carbones subministrarunt ardentes, evaporationi commisimus, et en! materiæ albæ salinæ tenuissimæ, drachmam unam cum duodecim granis obtinuimus.—Idem residuum in aquâ liquatum et per chartam colatum bibulam, præbuit lixivium, ex quo, lenioris exhalationis ope, alcalini salis scrupulos duos obtinuimus. *Hoffman de Element. et Viribus Fontis Selterani,* § V. Cap. II.

rated * with Fixible Air. I do not know in what quantity the Selters water is taken; but if it be drunk as largely as the Carlsbad water, about 150 grains, or two drachms and a half of the alkaline salt will be the daily dose, if we compute each *ollula* or glass at a quarter of a pint contents. Milk appears to be commonly joined with Selters † water in disorders of the kidneys, the milk of asses especially; and spices ‡ and other aromatics are also occasionally combined with it, much in the same manner as is advised in the present work.

Dr. Nooth, a gentleman whose knowledge and sagacity in philosophy, as well as in medicine, are well known, suggested to me the probability that the alkaline salt, saturated with Fixible Air, and crystallized, might

---

* Sicuti ergo singula hæc experimenta planum atque testatum faciunt fontem Selteranum affluentem salis alcalini puri copiam in complexu suo alere, ita pariter ex variis a me observatis phænomenis evictum esse credo, eundem instar omnium aquarum salubrium maximeque acidularum, purissimo illo volatili et minerali spiritu esse imbutum. Ibid. § VI. Cap. II.

Nec dubitare amplius licet, quin eidem spirituoso minerali ingredienti, salubris harum aquarum facultas quod maximam partem tribuenda est. Ibid. § IX, Cap. II.

† Ibidem, § XI. XII. XIII. XIV.

‡ Ibidem, § XIX.

might perhaps be equally effectual as a lithontriptic, and in abating the acrimony of the urine, as the mephitic alkaline solution, whilst it would be more convenient, as being easily portable, and also as it would be free of any stimulus, which the superabundant quantity of Fixible Air might produce on the urinary passages.

Not having made a trial of this preparation, I cannot answer for its effects in this way, but should think it not unlikely to succeed, especially when we reflect that a salt of a similar nature, crystallized from the Carlsbad waters, is recommended by Hoffman * in such complaints. If such a preparation

---

* Et cum ob abforbentem fuam facultatem, acidum, cujus cum terrâ connubium gignit calculum, cicurare et in medium fal convertere valeat; hinc tam ad avertendam tartarei coaguli genefin, quam exturbandum minorem calculum infignis eft efficaciæ. Quo nomine etiam iis commendari vix poteft, qui, vel ob nativam, vel ab alio qualicunque errore diætetico, contractam renum et urinariæ veficæ imbecillitatem, ejufmodi calculofis concretionibus et generandis et fovendis funt idonei. Quemadmodum pariter fingularum ejus experiuntur efficaciam, qui vehementes a calculo in arctum et nervofum ureterum cavum intrufo, dolores fentiunt, quibus, præ omnibus aliis medicinis lithontripticis, tam refolvendo, præfertim fi recens fuerit tartarei muci concrementum, quam exturbando illius copiam egregiè et citò, opitulatur fal Carolinarum. *De Sale Medicinali Carolinarum,* § XXII.

ration be attempted, it will be neceſſary to uſe the greateſt caution in previouſly ſaturating the alkaline ſolution with Fixible Air to its fulleſt extent, and of carrying on the evaporation as gradually, and of courſe with the leaſt heat poſſible, and frequently removing it from the fire, and ſuffering the ſalt to cryſtallize, which laſt proceſs alſo ſhould be gradual; and therefore, when the liquor is removed from the fire, it ſhould cool very ſlowly, and when cold be carefully poured off from the cryſtals that are formed, and again evaporated in like manner. I apprehend that it would be proper, when the liquor is evaporated pretty nearly to the point at which cryſtallization would begin to take place, when the liquor ſhould be cold, to put it again into the glaſs machine, and impregnate it a ſecond time with Fixible Air; which will probably, by ſupplying the Fixible Air that may have been diſſipated by the heat, accelerate the cryſtallization, and enſure the neutralization of the alkali. The mephitic alkaline ſolution, when meant to be evaporated, ſhould be made much ſtronger than when it is meant to be drunk, yet ſhould not be ſo much loaded as to impede or clog the tubes through which the Fixible Air riſes. About four ounces of ſalt of tartar to a pint of water, seems likely to prove a proper quantity. It should be noticed that the stronger the solution, the longer time it would require to be neutralized by the Fixible Air.*

Perhaps the fossil alkali might be more convenient for this purpose, as it crystallizes more easily than the vegetable; but it must be observed also, that the fossil alkali will crystallize before it be perfectly neutralized by the Fixible Air, and on that account more caution will be necessary in saturating it perfectly before any evaporation be commenced, and of conducting the evaporation itself as slowly as possible.

Experience will best ascertain the dose in which such a remedy may be taken. Perhaps one drachm daily might be sufficient to begin with, which might be gradually increased to two, three, or more. Hoffman says, that the Carlsbad salt is taken to six drachms, or an ounce, without producing any other effect than that of proving gently laxative.

To

* Should any person be inclined to make a trial of the Aerated Alkali, either the Vegetable or the Fossile, it may be had, ready prepared, in great perfection, of Mr. Thomas Willis, at the Hermitage, Wapping, a most ingenious practical chemist.

To what the wonderful propensity, in certain habits and constitutions of body, to generate urinary calculi, is owing, is yet undiscovered. Various modes of life, and regimens of diet, have been assigned as causes, and many facts have been adduced as proofs; but these accounts are all so ambiguous, inconsistent, and contradictory, that little can be concluded from them. Hard waters are at one time believed to produce them; at another, they rather tend to prevent their generation: wine is at one time preventive, and at another productive, of calculi; and malt liquor, which some condemn, is by others as extravagantly recommended*.

It appears highly probable, that the cause originally consists in the structure or nature of the secreting gland itself. By what means this can change the quality of the fluid, so as to render it at one time apt to precipitate its contents, and at another to hold them in perfect solution, is to us inconceivable; but not more so than the generation of blood from the chyle, or of bile from the blood, the mechanism or process of which is probably among the secrets of nature too deep for our comprehension. It is at least certain, that confinement to a certain posture will in
some

* See Medical Commentary, p. 128.

some instances produce this complaint. I have seen it originate from sitting long at a sedentary employment, as writing; and from long confinement to bed, by an illness no ways connected with calculus. Ramazzini makes the same observation of persons whose way of life requires a standing posture; which he instances by that of the attendants at the courts of princes, those of Spain especially, among whom disorders of this kind were particularly frequent.

Wether these theories be true or false; whether the remedy acts by means of the chemical combination with the fluid secreted, or by any still more obscure means upon the secreting organ itself; the facts still remain unimpeached. The cases above related evince, beyond a doubt, that the painful symptoms in calculous disorders have been removed, and ease procured, by the use of this remedy, and this without any ill effect on the general health; but, on the contrary, with great amendment of it in most cases. To account for these, is the province of philosophical investigation; and with that view I have, however imperfect they may be, offered my sentiments: but whether the opinion I have here adopted be well or ill founded, the facts are equally

equally valuable, and will, I truſt, encourage the farther trial of a remedy, which, in a manner the moſt eaſy, and favourable to the health in general, bids fair to relieve, in a degree hitherto unexperienced, one of the moſt excruciating diſorders that is incident to human nature.

POSTSCRIPT.

THE following Caſe, which did not arrive until the work was ſent to preſs, is too important to be omitted, as it points out the uſe of this remedy in a complaint of the urinary paſſages, unattended with calculus, and probably owing to a peculiar acrimony in the ſecretion itſelf. This caſe is atteſted by a gentleman of great eminence in his profeſſion, and whoſe candour in the narrative does him the higheſt honour.

A young woman in her 29th year, and who had hitherto enjoyed very good health, began in the month of March, 1789, to have frequent

frequent dull pains in the region of the bladder, and to paſs her urine frequently, and in ſmall quantities, attended with acute pain and ſymptoms of ſtone. On this account ſhe was founded; and no ſtone being found, her complaint was ſuppoſed to be occaſioned by a ſchirrhus at the neck of the bladder, and this opinion was ſtrengthened by her always deſcribing a ſenſe of weight there. The weather being unfavourable for a mercurial courſe, it was thought expedient to give her the Cicuta till the ſummer advanced; and that medicine was perſiſted in for two months without any relief: her ſtomach then began to reject it, even in the ſmalleſt doſe; and it conſequently was diſcontinued. She now was confined to a regimen of milk, farinacea, and marſhmallow tea, in which gum arabic was diſſolved: for a few days ſhe thought herſelf rather better; but at the end of a fortnight a new appearance took place, ſhe began to paſs large quantities of mucus with her urine; and from this period her pain increaſed to that degree as to require the occaſional uſe of large doſes of opium. In the middle of July ſhe began a mercurial courſe, and ſubſtituted a pill of the Extractum Hyoſcyami at bedtime for one of opium; which was continued to October, without producing any other change

change than a great diminution of strength. Being now tired of medicine, she requested to desist from every thing of the kind, except the pill with opium, which *alone* afforded a temporary suspension of pain. In the middle of October she went to pass the winter with her friends in the south, and did not return till the May following. The appearance in her urine was now changed: instead of large quantities of mucus, she passed little or none; and once in four or five days she evacuated bloody urine; and this evacuation was constantly preceded by lancinating pains and a sense of weight at the neck of the bladder; which sensation left her after the discharge took place, and she then remained tolerably easy for a day or two. The disorder now appeared very analogous to the piles: and Sauvages, in his Nosologia Methodica, under the title of Hæmaturia Hæmorrhoidalis, refers to apparently similar cases, noticed by practical writers. On account of this new symptom, she was directed to take small doses of the vitriolic acid, by taking two ounces of the tincture of roses every four hours: but this medicine, after a short trial, producing no effect, she requested to take the Mephitic Alkaline Water, which had been strongly recommended to her by a friend, who had
experienced

experienced very great relief from it in a case of gravel, producing occasionally bloody urine. On considering the various means that had been hitherto unsuccessfully employed for her recovery, it was thought advisable to consent to her request, although, upon the strictest examination of her urine from time to time, not the smallest particle of sand or gravel had ever been discovered: accordingly she began to take it as directed in the Treatise on Aqua Mephitica Alkalina, by Dr. Falconer; and in a few days she was sensible of a great abatement of pain, and some decrease in the appearance of blood in her urine; and thus she continued gradually to recover, and in six or seven weeks became perfectly well, and has continued so, notwithstanding she has left off the use of the water for some time.

WILLIAM INGHAM.

Newcastle, Dec. 21, 1791.

The following Case came too late for insertion, unless where it is now placed.

It is a notable instance of the lithontriptic powers of the Aqua Mephitica Alkalina. It is highly probable that the great number of of fragments which the patient voided in December laſt, were once concreted into one or more large calculi, and that the diſſolvent power of the remedy ſeparated them into portions, ſo ſmall as to admit of being diſcharged by the urinary paſſages. The ſoftneſs of conſiſtence of the laſt which he paſſed ſeems to put this ſuppoſition nearly beyond a doubt, and to eſtabliſh this quality of the remedy on the moſt reaſonable grounds.

Norwich, 24*th February*, 1792.

Mr. G. Harwood, an eminent attorney of this place, had for ſome years been troubled with ſuch complaints as clearly indicated either the retention of ſabulous matter in the kidneys and bladder, or the formation of a calculus in the latter. About the midſummer of the year 1789 theſe complaints were much increaſed, and he was recommended to try the Alkaline Mephitic Solution. After he had taken this rather more than twelve months he paſſed eight ſtones, all with ſmooth ſurfaces, the largeſt of theſe about the ſize of a common pea: from this time he regularly perſevered in the uſe of the ſolution, and in December laſt was ſeized with a moſt violent

violent attack; and in the courfe of rather more than a week he voided nearly one hundred and fifty pieces of ftone, and at laft one fmall ftone fo foft that it would have broken to pieces with the flighteft preffure.

Since this time he has had no return of his complaint; nor has he made any bloody urine (which before this the gentleft exercife ufed to promote), although he fometimes walks three miles.

He ftill continues to take the folution; as, before the laft attack, he ufed at times to void fmall *pieces* of ftone, *but no whole one*, which he altogether attributes to its ufe.

    W<small>M</small>. ATTHILL, S<small>URGEON</small>.
G<small>ARD</small>. H<small>ARWOOD</small>.

F I N I S.

www.ingramcontent.com/pod-product-compliance
Lightning Source LLC
Chambersburg PA
CBHW020831230426
43666CB00007B/1184